Multimedia Engineering

RSP SERIES IN INDUSTRIAL CONTROL, COMPUTERS AND COMMUNICATIONS

Series Editor: **Professor Derek R. Wilson**

Concurrent Engineering – The Agenda for Success
Edited by **Sa'ad Medhat**

Analysis and Testing of Distributed Software Applications
HENRYK KRAWCZYK **AND** BOGDAN WISZNIEWSKI

Interface Technology: The Leading Edge
Edited by **Janet M. Noyes** and **Malcolm Cook**

CANopen Implementation: Applications to Industrial Networks
Mohammad Farsi and **Manuel Bernado Martins Barbosa**

J: The Natural Language for Analytic Computing
Norman Thomson

Digital Signal Processing: A MATLAB-Based Tutorial Approach
John Leis

Mathematical Computing in J: Introduction, Volume 1
Howard A. Peelle

System Building with APL + Win
Ajay Askoolum

Multimedia Engineering: A Practical Guide for Internet Implementation
A. C. M. Fong and S. C. Hui, with contributions from G. Hong and B. Fong

Multimedia Engineering

A Practical Guide for Internet Implementation

A. C. M. Fong
Nanyang Technological University, Singapore

S. C. Hui
Nanyang Technological University, Singapore

With contributions from G. Hong and B. Fong

John Wiley & Sons, Ltd **Research Studies Press Limited**

Published by John Wiley & Sons, Ltd., The Atrium, Southern Gate, Chichester,
West Sussex PO19 8SQ, England
Telephone (+44) 1243 779777

Email (for orders and customer service enquiries): cs-books@wiley.co.uk
Visit our Home Page on www.wiley.com

This Work is a co-publication between Research Studies Press Limited and John Wiley & Sons, Ltd.

Other Wiley Editorial Offices

John Wiley & Sons Inc., 111 River Street, Hoboken, NJ 07030, USA

Jossey-Bass, 989 Market Street, San Francisco, CA 94103-1741, USA

Wiley-VCH Verlag GmbH, Boschstr. 12, D-69469 Weinheim, Germany

John Wiley & Sons Australia Ltd, 42 McDougall Street, Milton, Queensland 4064, Australia

John Wiley & Sons (Asia) Pte Ltd, 2 Clementi Loop #02-01, Jin Xing Distripark, Singapore 129809

John Wiley & Sons Canada Ltd, 22 Worcester Road, Etobicoke, Ontario, Canada M9W 1L1

Wiley also publishes its books in a variety of electronic formats. Some content that appears in print may not be available in electronic books.

Library of Congress Cataloging-in-Publication Data

Fong, A.C.M.

Multimedia engineering: A practical guide for internet implementation / A.C.M. Fong, S.C. Hui; with contributions from G. Hong and B. Fong.

p. cm.

Includes bibliographical references and index.

ISBN-13 978-0-470-03019-6 (cloth)

ISBN-10 0-470-03019-4 (cloth)

1. Multimedia systems. 2. Internet. I. Hui, S. C. (Siu Cheung). II. Title: A practical guide for internet implementation

QA76.575.F664 2006

006.7--dc22

2006005399

British Library Cataloguing in Publication Data

A catalogue record for this book is available from the British Library

ISBN-13 978-0-470-03019-6 (HB)
ISBN-10 0-470-03019-4 (HB)

Typeset in 10/12pt Times New Roman by Laserwords Private Limited, Chennai, India
Printed and bound in Great Britain by TJ International, Padstow, Cornwall
This book is printed on acid-free paper responsibly manufactured from sustainable forestry
in which at least two trees are planted for each one used for paper production.

TABLE OF CONTENTS

PREFACE

The field of Information Engineering (or Information Technology (IT)) has emerged from the more traditional disciplines of Electrical Engineering and Computer Science. Many have compared the current IT developments to the industrial revolution, which introduced the factory system and brought widespread manufacturing to the world.

IT has certainly had an impact on many facets of modern life, arguably in a way that has never happened before in terms of its speed of coverage and geographic spread. In a mere decade or so, IT has revolutionized the way people, work, communicate and search for relevant information. IT is now applied to such diverse areas as commerce, industry, government, law enforcement, entertainment, education, health care, and so on. The list is endless. Many would argue that this process of revolution has only just begun, with further convergence of both technology and applications being the goal on the near horizon, beyond that nobody knows for sure. Even the term "convergence" is constantly being refined. To many people, this means combining voice and data traffic, wireless and wired networks, and so on, through which wonderful things will happen to improve our quality of life. Already, we have technologies like "click-and-talk" web pages, unified messaging and multimedia conferencing, all of which offer both opportunities and challenges.

At the heart of this revolutionary process are advancements in a number of enabling technologies. Two key elements are the Internet (especially the World Wide Web) and Multimedia. The primary purpose of this book is to focus on the recent innovations in these two key elements, which underpin the explosive growth of IT applications.

We have written this book to focus on IT applications and therefore we do not present fundamental theory such as sampling, digital-analog conversion, quantization, programming, computer architecture, and so on. We also assume basic programming skills using C/C + + and Java, as well as common Internet terminology such as Uniform resource locator (URL). All these topics can be found in traditional Electrical Engineering or Computer Science textbooks. Nor is the emphasis on the underlying networking aspects of the Internet. As this book is written, broadband access is rapidly increasing, and will continue to do so.

Based on our own teaching experience, we have identified that there is a need for a book that focuses on recent advancements in the Internet (especially the World Wide Web) and the applications of multimedia technologies. Consequently, this book provides surveys of recent solutions and implementation technologies. It is a culmination of years of research that have identified shortcomings in existing solutions, which has led to the development of novel solutions to engineering problems and working systems that represent advancements in the state-of-the-art. The reader is provided with a unique insight into the developments of working systems in numerous "walk-through" problem-solution examples.

This approach differentiates this book from most others in the treatment of the subject. We have intentionally written the book so that each chapter is self-contained. So, for instance, one can read Chapter 3 on using the Internet for communications without spending any time on Chapter 2. This allows a mix-and-match approach of topic selection, which is particularly suitable for classroom instruction purposes. It also facilitates the self-learning process in that readers can select the material they "NEED TO KNOW", without having to backtrack to other material.

Writing this book has been a significant intellectual challenge in that we have striven to present advances in the technology within an application context. We have enjoyed this challenge and we hope you will enjoy the book and find the material useful.

A. C. M. Fong
S. C. Hui

ACKNOWLEDGEMENTS

The authors would like to thank Prof. Derek Wilson, Series Editor, for coming up with the idea of this book and for his encouragement and valuable comments throughout its development. Giorgio Martinelli at Research Studies Press also deserves much credit for making this book a reality. He has been most helpful and patient throughout this project. The authors would also like to thank all others who have helped in any way, especially the two contributing authors, G. Hong and B. Fong, as well as those who have contributed to the projects described in this book. These include Yulan He, Ho Le Vu, K. V. Chin, L. S. K. Chong, G. Jha and Pui Y. Lee.

A. C. M. Fong
S. C. Hui

CHAPTER 1

THE DAWN OF A NEW AGE — THE INFORMATION AGE

1.1 THE INFORMATION AGE AND THIS BOOK

In this Information Age, information is power, and is often considered an engine for future growth and prosperity. It has also created an environment where small companies and developing countries can compete and catch up more easily than ever before. Not since the industrial revolution has humankind witnessed such an explosive rate of change in which every aspect of our lives is affected by the current information technology (IT) revolution.

This IT revolution has caught on like a wildfire, reaching every corner of the world. It would not be an exaggeration to say that the IT revolution has touched our lives in every way imaginable, from commerce, military, service and manufacturing industries, through to government, law enforcement, entertainment, scientific pursuits, space exploration, and so on. Not only is this list seemingly endless, it is still growing. Despite all the recent advancements in information technology, many believe that this revolution has only just begun. We are but at the dawn of this IT evolution. While it is unclear where all this will lead us, a relative certainty is that advancements in Internet and multimedia technologies, which have contributed tremendously to this IT revolution, will continue to play an important role in this revolution in the foreseeable future.

It is for this reason that we have prepared this book, the purpose of which is to provide the reader with a timely and relevant sample of recent advances in Internet and multimedia technologies at the dawn of this IT revolution. It is but a snapshot of a fast growing and dynamic field of research, which is very much multidisciplinary in nature. We have deliberately left out fundamental concepts (such as basic digital communications, computer architecture, signal processing, programming, etc.) that can be found in any good electrical engineering or computer science textbook. Rather, we assume that the reader already has a sound background in these

Multimedia Engineering A. C. M. Fong & S. C. Hui

topics on which our work is built. Thus, the reader will not find material such as Fourier transform, sampling, quantization, and so on in this book. Instead, each chapter generally follows a survey-problem-solution format. It is intended to give the reader a first person view of why and how certain engineering solutions are arrived at and what problems are being solved by those solutions. In each case, a performance analysis is also presented where appropriate.

The remainder of this introductory chapter serves two purposes. The next section defines the three important terms that form the basis of this book, namely, the Internet, World Wide Web and multimedia. This is followed by a section that summarizes the subsequent chapters, which are categorized into four major parts, namely, fundamental concepts, applications, software implementation issues, and wireless deployment. With advances in computing and communications technologies, not to mention the much-touted convergence of Internet and wireless networks, we anticipate an increase in data traffic on heterogeneous networks that are made up of both wireline and wireless networks.

1.2 THE INTERNET, WORLD WIDE WEB AND MULTIMEDIA

Until about a decade ago, terms such as the Internet, World Wide Web and Multimedia were largely unknown outside the scientific community. Today, these words are widely known throughout the world, and are known to people of all ages from 3 to 80 (or beyond). We know that technological advancements in these areas have contributed largely to the phenomenon that we describe as the dawn of the information age, but what do these terms mean exactly? Since this is what this book is all about, it is important to clearly define these terms from the outset.

1.2.1 The Internet

In a nutshell, the Internet is a huge interconnected set of disparate networks. The components (or sub-networks) that make up the Internet can (and do) have very different characteristics, such as bandwidth resources, the degree of Quality of Service (QoS) Support provided, and so on. All these make it both challenging and interesting to make effective use of the Internet. To really appreciate what the Internet is all about, it is best to review a brief history of it.

1.2.1.1 A Brief History of the Internet and Related Networks

It is generally accepted that the predecessor of what is now known as the Internet began life in the late 1960s/early 1970s when the United States Defense Advanced Research Projects Agency (DARPA) initiated research into ways of interconnecting packet networks of different kinds. A major thrust of the research was to develop suitable communication protocols to enable the transfer of packetized data over the interconnected networks. The word “Internet” emerged to describe the interconnected networks and the fundamental protocol became known as Internet protocol (IP). Closely related to IP is the Transmission Control Protocol (TCP), which, along with IP, formed the well-known TCP/IP Protocol Suite that is still frequently used today.

In the mid 1980s, the United States National Science Foundation (NSF) began the development of National Science Foundation Network (NSFNET), which currently provides a major backbone communication service for the Internet. In addition, the National Aeronautics and Space Administration (NASA) and the United States Department of Energy (DE) also support the Internet backbone with their NASA Science Internet (NSINET) and Energy Sciences Network (ESNET), respectively. Today, the Internet backbone is also supported in Europe and elsewhere. On a local scale, individual research and educational institutions such as universities and research laboratories support the Internet.

By the late 1980s, the number of individual implementations of the TCP/IP protocol suite developed by both public institutions and private organizations and companies had reached approximately one hundred. The Internet has continued to evolve and by the early 1990s it became apparent that integration with other protocol suites would be an important issue. In particular, there has been an ongoing effort to ensure multi-protocol internetworking, especially the integration of the Open Systems Interconnection (OSI) protocols into the Internet architecture. OSI protocol implementations also became available. By the end of 1991, the Internet has grown to include approximately 5,000 networks in about forty countries, serving over 700,000 host computers used by over 4,000,000 people.

Over the course of the Internet evolution, financial support has shifted from government-funded research (especially by the United States government) to privately funded entities and other governmental agencies around the world. The Internet is now a truly international phenomenon.

1.2.1.2 Custodians of the Internet

Today, the Internet is largely made up of private networks in educational and research institutions, businesses and in government organizations around the world. With so many participants, it is necessary to instill effective coordination to oversee the evolution of the Internet. The United States Federal Networking Council (FNC) and the European Reseaux Associees pour la Recherche Europeenne (RARE) have jointly set up the Coordinating Committee for Intercontinental Networks (CCIRN) to coordinate government-supported research activities and to promote international cooperation in the Internet environment.

In terms of protocol developments, the Internet Activities Board (IAB) was established in 1983 to guide the evolution of the TCP/IP Protocol Suite and to provide technical advice to the Internet community. After a few rounds of reorganization, the IAB today has two main functions: the Internet Engineering Task Force (IETF) and the Internet Research Task Force (IRTF) The former has primary responsibility for further evolution of the TCP/IP protocol suite, its standardization with the concurrence of the IAB, and the integration of other protocols into Internet operation (especially OSI protocols). The Internet Research Task Force continues to organize and explore advanced concepts in networking under the guidance of the IAB and with support from various government agencies.

In addition, the IAB also publishes Internet-related technical documents and records identifiers for protocol operations. Initially, Internet and protocol-related documents were published in a series known as Internet Experiment Notes (IEN). Subsequently, the documents are published in a series known as Requests for Comment (RFCs), which are still in use today. The first RFCs were used to describe the protocols of The Advanced Research Projects Agency Network (ARPANET), the first packet switching network developed by DARPA in 1969.

The identifiers for protocol operations are recorded by the Internet Assigned Numbers Authority (IANA), especially the Internet Registry (IR), which serves as a central repository for Internet information. The IR provides a central allocation of network and autonomous system identifiers, and in some cases to subsidiary registries located in various countries. The IR also provides central maintenance of the Domain Name System (DNS) root database which points to subsidiary distributed DNS servers replicated throughout the Internet. The purpose of the DNS distributed database is to associate host and network names with their Internet addresses. This is critical to the operation of the higher-level TCP/IP protocols including electronic mail.

There are also Network Information Centers (NICs) located throughout the Internet to serve its users with documentation, guidance, advice and assistance. As the Internet continues to grow, the need for high-quality NIC functions increases.

1.2.1.3 Other Networks

The early 1980s also saw the initiation of two networks: Because It's Time Network (BITNET) and Computer Science Network (CSNET). The BITNET was originally developed for use in the academic community with a multidisciplinary flavour. It adopted the International Business Machines Corporation (IBM) Remote Spooling Communications Subsystem Networking (RSCS) protocol suite using direct leased line connections between participating sites, which now number several thousands worldwide. Recently, BITNET has established a TCP/IP backbone with RSCS-based applications running above TCP.

In contrast to the multidisciplinary use of BITNET, CSNET was developed to interconnect computer science research groups based in universities, the industry and government organizations. In 1987, BITNET and CSNET merged to form the Corporation for Research and Educational Networking (CREN). However, CSNET service was terminated in 1991 after serving its intended purpose. At its peak, CSNET provided connections for about two hundred participants in over a dozen countries. Today, funding for the CREN is provided by the participating organizations.

1.2.1.4 The Success of the Internet

One of the strongest points of the Internet is that with the exception of real time (multimedia) applications, it makes use of marginal resources. This means using network resources that might otherwise be wasted, and is achieved by breaking data down into packets and sending the data via whatever route has capacity, without giving much regard to issues such as latency. This contrasts with a circuit-switched scenario, notably the Public Switched Telephone Network (PSTN), which locks up valuable resources for the duration of a data transfer session (e.g. a telephone conversation). Internet traffic is therefore an efficient way to share resources among uses.

In addition, the real costs of computing power and data storage have continued to decline. At the same time, there is also a shift towards mobile computing as portable computers become smaller and lighter, some of which can even be integrated into clothing articles to become wearable computers. Further, many sources predict that Internet access charges will continue to come down, as in [1]. All these factors that have fuelled the phenomenal success of the Internet are set to continue to do so in the foreseeable future.

1.2.2 The World Wide Web

The World Wide Web (or simply WWW or Web) can be considered both a cause and an effect of the information age. It is very much a product of the information age, which began in earnest about a decade ago, and yet it is very much a facilitator of further development at this dawn of the information age. The Web is very simple to use, and yet it can be very powerful. But what exactly is the Web? To answer this question, we begin by taking a brief look at the Web's history.

1.2.2.1 A Brief History of the Web

It is generally accepted that the Web is attributed to Tim Berners-Lee (who has become the Director of the World Wide Web Consortium or W3C) and other physicists working at the European Particle Physics Laboratory (CERN) in Geneva. From the outset, the Web was developed for solving a particular problem faced by scientists at different locations. Physicists working at CERN and elsewhere wanted a means to share information, such as the latest theoretical developments and experimental data. They wanted to do so rapidly and to include participants from thousands of miles away. They also wanted to include multimedia data comprising a mix of text, images, and so on.

Thus, the Web was born with the publication of a proposal [2]. At a high level of abstraction, it was envisaged as a scalable and dynamic structure comprising a connected web whose interconnections change with time. Within this conceptually huge structure, the first problem that had to be solved was one of navigation. In 1992, the CERN scientists came up with the first working Web browser and the associated Web server program. A Web browser provides a means by which a user can request and view documents posted on the Web. Although the original CERN Web browser was rather rudimentary, it provided a working solution to prove the feasibility of the project. Today, popular Web browsers such as Netscape Navigator and Microsoft Internet Explorer have become commonplace and are supported by different computer platforms.

While the Web browser serves as an interface between the human user and the user's computer, a Web server program serves as an interface between the user's computer and the Internet. Its primary purpose is to connect the user's computer to the Internet, and at the same time to offer its services to other Internet users as appropriate. More specifically, when the user requests a Web document via the Web browser, the Web server receives and processes the request. It then relays the requested document from the Internet to the Web browser for the user. Typically, the user may choose to display the document on the Web browser or to store it onto the local hard disc for subsequent use.

1.2.2.2 The Strengths of the Web and Supporting Technologies

Much of the success of the Web is attributed to the supporting technologies that facilitate the exchange of information easily and on a truly global scale. From the users' perspective, the Web provides an ideal forum to exchange information with equipment that an increasing number of individuals can afford throughout the world. All a user needs is a personal computer (PC) and an internet connection. The Web enables users to disseminate information across the globe with minimal know-how needed.

Perhaps the most significant factor that has contributed to the success of the Web is the idea of modelling the Web as a loosely interconnected network of resource providing servers that make up the Internet. This not only means individual content providers can take complete ownership of their contents; the storage requirements of all contents on the Web are thus distributed throughout the Web. Individual content providers, therefore, rely on themselves to provide the necessary servers and storage systems for their own contents. This makes the Web highly scalable.

Other factors that have significantly contributed to the success of the Web include markup language, multimedia contents, interactivity, the Java programming language and the universal resource locator. The first, and still most widely used markup language is the Hypertext Markup Language (HTML). HTML and the more recent markup languages (such as Extensible Markup Language - XML) are characterized by being widely used and simple to use. These markup languages provide a simple means to format Web documents, as well as a simple mechanism for linking Web documents. The documents can also include multimedia contents that enhance the presentation of the material. In addition, a Web page can easily be made active, in such a way that a downloaded program can respond to user inputs almost instantaneously without having to execute the program on the Web server side every time the user makes an input.

The interoperability/platform-independent characteristics of Java allows programs to be downloaded onto the user's computer to support, for instance, the development of active Web pages described above. Most importantly, the content provider does not need to make specific reference to the target computer platform when Java is used. Since the inception of Java by Sun Microsystems in 1995, it has been widely adopted by major players in the computer industry, such as Netscape, which was the most widely used Web browser company at the time.

The universal resource locator (URL) is essentially a simple addressing mechanism for identifying Web documents. It includes information such as the location of the host server on the Internet, the location of the contents within the host, and the type of contents (e.g. html documents). The URL provides a simple — and often meaningful — means to uniquely identify a Web document. In fact, a URL does not only just identify a Web document, it leads the user to the actual document identified by the URL in question. For example, the URL http://www.ntu.edu.sg uniquely identifies the home page of Nanyang Technological University. By inputting this URL in the appropriate field of any Web browser, the home page is displayed on the user's computer screen.

1.2.3 Multimedia

In the context of information technology, a medium is a means by which we represent information. For example, information may be represented in textual format, such as the words in this book. Information may also be represented in other forms, such as still images, video (moving pictures), audio (speech, music, soundtracks, etc.). In a broader sense, the information presented may not even be immediately comprehensible by a human reader/observer/reader. Information may also be encapsulated in the form of data recognizable by computers.

In a nutshell, therefore, multimedia can be thought of as a term that collectively means a mix of different media, or representations of information, whether or not the representations are immediately comprehensible by human beings. In addition, it is generally accepted that multimedia representations should be taken to mean digital signals. This means speech signal or music stored on analog tape is generally not considered multimedia. Thus, in the information age, multimedia is taken to mean a mix of media for information representation (text, image, video, audio and computer data) in digital format.

However, multimedia information is of little value by itself. Provisions must also be made to manipulate, store and transmit the multimedia information. For example, it is possible to enhance the quality of digital images by applying image processing techniques such as contrast enhancement or histogram equalization. Multimedia information can be stored on the local Web servers of individual content providers. The Internet can also be considered a vast networked communications channel for the transmission of multimedia information. Thus, in the broadest sense, the word multimedia encompasses not just the media (or representations) themselves, but also the methods of manipulation, storage and transmission of the media. An Internet videoconferencing system is a good example of multimedia systems. A sequence of images and speech data (of the participants) are processed, stored (or buffered) and transmitted from one node of the Internet to others.

1.3 ORGANIZATION

From the discussions presented in Section 1.2 above, the basic theme of this book emerges. The fundamental contribution of the Internet and multimedia to our society has been the dissemination of information, often in mixed modes of media comprising text, video, audio, and so on. Dissemination of information takes two primary forms. The first involves putting up information for other users to retrieve, in much the same way that a library provides books and other reference material for readers to seek information. The second form is a more direct human-to-human communication, such as electronic mail, Internet telephony and videoconferencing.

Chapters 2 and 3 present these two fundamental forms, which provides the platform on which many applications, such as electronic commerce, are built. In particular, Chapter 2 considers the Internet as a vast information repository. The challenges involve effective storage, indexing and retrieval of text and other data formats. Topics that are covered in depth in Chapter 2 include search engine technologies, personalized monitoring services, storage and retrieval of visual data, and a case study focusing on the discovery and monitoring of scientific publications available on the Web.

Chapter 3 considers the Internet from the perspective of using it as a communications medium, which offers tremendous potential for low cost human-to-human communications (e.g. Internet phone as compared to traditional circuit-switched PSTN). However, the current best-effort Internet is a harsh environment in which to transmit data. This chapter, therefore, describes methods of overcoming the network congestion problems so that the potential of the Internet as a communications channel can be fully exploited. This chapter begins with a discussion of the various protocols for Internet communications. In-depth topics such as electronic mail systems, instant messaging systems, Internet telephony, video data transmission, desktop videoconferencing and unified messaging.

Before useful applications can be built, issues regarding Internet security and privacy have to be resolved. Chapters 4 and 5 present the different facets of Internet security and privacy issues. In particular, Chapter 4 presents topics on Internet security and practical approaches, security challenges for Java-based systems (using an Internet Java phone as an example), multi-view facial analysis as a means of biometric authentication. Privacy is an important issue, both for private individual users and enterprises. Chapter 5 focuses on effective Web content filtering applicable to both groups of Internet users.

A variety of industrial and commercial applications of the Internet and multimedia are presented in Chapter 6. Topics presented in Chapter 6 include a virtual trading system for business-to-business (b2b) electronic commerce, Web-based customer technical support, knowledge discovery for managerial decisions, and Web-based security surveillance.

Chapters 7 and 8 concern the development and deployment of practical Internet and multimedia systems. Chapter 7 (by G.Y. Hong) then presents the latest trends in the development of reliable practical Web-based systems from a software engineering perspective. With technological advancements in wireless communications, mobile (m-)commerce will be become a reality for many business transactions in the near future. However, a particular challenge for supporting m-commerce is the development of highly reliable wireless communications. To support business transactions for mobile users, the level of system reliability must be much higher than for traditional applications such as mobile telephony that carry only speech data. Chapter 8 (by B. Fong) describes some of the latest research findings associated with the deployment of wireless systems suitable for m-commerce applications.

References, Links and Bibliography

[1] F. McInerney and S. White 'The internet's sustainable advantage', *IEEE Computer*, Vol. 30, No. 6, pp. 118–120, June 1997.
[2] http://www.w3c.org/history/1989/proposal.html, 2004.

CHAPTER 2

THE INTERNET AS AN INFORMATION REPOSITORY

2.1 INTRODUCTION

The Internet is fast becoming a popular medium for sharing information due to its continuous availability and wide geographic coverage. It is easy to understand why the Internet appeals to the masses as an information dissemination medium of choice. Nowadays, anyone with little computing skill and affordable equipment can easily make information available on the Internet for others, who also do not require sophisticated skills or equipment, to access it. After all, the Internet traces its roots back to the ARPANET (Advanced Research Projects Agency Network), which was as much a communication medium as it was a means of sharing information.

Virtually everything about anything is now available on the Internet (or more specifically, the Web). Ironically, with the wealth of information on the Web, many users are experiencing the effects of information overload — they are simply overwhelmed by the amount of information available. This makes finding relevant information difficult.

With an estimated 500–800 million Web pages, coupled with an explosive growth at one million pages a day, the likelihood of finding something useful is diminishing. It is becoming increasingly difficult to sift through the information to find what is desired, or what is most closely matched with what the user is precisely looking for. Search engines such as those provided by Google [1], Yahoo! [2], Excite [3], Lycos [4], and so on, have been developed to alleviate the effort involved in finding useful information on the Web. However, most users know that while these search engines are very useful, there is still much to be desired in terms of their performances.

In this chapter, we consider various aspects of using the Internet (specifically the Web) as a huge and widely accessible information repository. In particular, we shall examine the following:

Multimedia Engineering A. C. M. Fong & S. C. Hui

- The promises and challenges of using the Web for information dissemination and retrieval.
- Search engines and so-called *meta search engine*s, as well as their underlying principles in terms of the indexing and searching approaches employed. These approaches rely heavily on hyperlink and Web page analysis.
- Personalized monitoring services.
- Indexing and retrieval of image and video data.
- A case study focusing on the discovery and monitoring of Web scientific publications.
- Further advancements.

2.2 CURRENT STATUS, PROMISES AND CHALLENGES

The amount of information currently available on the Web is staggering, as is the diversity of the kind of information available. Anything from general information such as the latest news and current affairs, weather, finance, sports, recipes, history, telephone directories, maps, information on products and services, and so on, through to information of a more specialist nature such as patents, scientific publications and citations are now readily available, literally at one's fingertip. A few keystrokes are all that is necessary to find the required information, from astronomy to zoology.

At least, this is the theory. Many users are not overly concerned that they eventually find the desired information, albeit after several rounds of trying and possibly after using a number of different methods (guessing the URL [Universal Resource Locator] of the wanted Website, asking others either online or offline and/or using a number of search engines, trying a number of different search strings along the way).

The much-touted convergence of mobile telephony and personal computing promises to bring Internet usage to a new level. This convergence is brought about by high capacity wireless transmission technologies, availability of portable and powerful computing devices and integration of mobile phones with some form of computing capabilities. With all these advances in mobile computing technologies, more and more users are accessing the Internet on the move, away from the traditional access points in their offices, schools and homes. For example, Microsoft and Bosch have announced a new initiative to bring the Internet to the automobile [5]. It is designed to combine Internet connectivity with traditional car navigation and route-planning functions, as well as entertainment systems for drivers and passengers. When combined with the global positioning system (GPS), up-to-the-minute driving directions/conditions can be provided to the driver accurately and virtually instantaneously.

As the number of Web pages continues to soar, keeping track of (or indexing) all available Web pages becomes increasingly difficult, if not impossible. This is compounded by the fact that most Web pages are designed to be dynamic. Indeed, one of the major attractions of publishing information on the Web is to be able to update the contents easily and as often as necessary. Moreover, although the use of a mark-up language such as Hypertext Markup Language (HTML) and Extensible Markup Language (XML) provides some uniformity in terms of the syntax of each Web page, information on the Web as a whole is not very well organized. This means that once a relevant page is found, finding all related relevant pages is not necessarily a simple task.

It is somewhat ironic that it should be difficult to find relevant information on the Web when there is so much information available. Apart from a lack of organization among information posted on the Web, there are other factors that influence the accuracy and speed of information retrieval from the Internet. The first is a difficulty of injecting semantic meaning into Web contents by the publisher. This subsequently leads to difficulty experienced by end-users in formulating effective search queries. In addition, the quality, treatment and presentation of the information provided on the same topic tends to differ from Web page to page. For example, when a user is looking for a particular topic such as "fractals in image processing", it is very likely that the user will find some Web pages more informative than others, even though the different pages may be equally relevant. How can a user find information that is both relevant and of high quality? We shall look at ways of tackling these problems in the next section on Search Engines.

The dynamic nature of the information available on the Web provides an opportunity for tracking relevant information over a period of time. For example, one may be interested in following developments in certain news items, celebrities, technologies, and so on. Using the same search engine to search for a particular query term six months apart, say, is likely to yield different results. In Section 2.4, we describe a Web monitoring system that can be customized according to the needs of individual users.

Much of the advancements in Web content dissemination and retrieval have been reported for textual information. Retrieval of images and videos are becoming increasingly important as more Internet users are switching to broadband access, allowing them to download large files with relatively little latency. Retrieval of these visual items using visual information requires different approaches from those for textual information. Section 2.5 provides a survey of the current techniques and systems for such purpose.

Various Internet-based information systems have been developed and deployed in various domains for solving specific problems. While search engines are very popular Internet utilities, they can be considered a special instance of a more general (emerging) field of research known as *information retrieval* (IR). Effective access to a digital library is yet another instance of research in IR. Section 2.6 presents a case study of the discovery and monitoring of Web scientific publications.

Much has been accomplished over the past 10 years or so, since the Internet evolved into its present form as the preferred choice for the masses to publish — and seek — information, particularly on the Web. However, much more remains to be done to fully realize its potential as an information repository. Section 2.7 ends this chapter by highlighting research directions, such as semantic Web and human-centric query processing, which are expected to benefit Internet users soon.

2.3 SEARCH ENGINES

Search engines are among the most commonly used and powerful tools available to Internet users. Best of all, virtually all of the (general nature) search engines are provided free of charge. In return, search engine purveyors would like as many users as possible to use their search engines (preferably as home pages) since they primarily derive their revenues from advertisements placed on their Web pages. This fundamental economical consideration provides a strong incentive for the different search engine purveyors to develop good search engines.

There are said to be over 150 commonly used search engines available. While most of these such as Google, Yahoo!, Excite and Lycos are general-purpose search engines, some are designed to cater for specific needs. For example, Deja News [6] is an online archive of all Usenet groups dating back to March 1995. Usenet postings are characterized by being short-lived, most of which are erased within a fortnight according to Deja News. Like most search engines, Deja News allows query filtering by date, Usenet group, author, subject or the appearance of keywords. On the other hand, it also provides an author profile with statistics on the number of posts made, and to which groups, from a particular e-mail address.

The primary focus of this section is on general-purpose search engines. This is an important topic, not just because search engines are very popular in their own right, but also because some of the specialist applications also rely on general-purpose search engines to perform some of the required tasks. This section is not intended to provide a rundown of the features of the various search engines. Rather, the focus is on the underlying techniques. The emphasis here is on the engineering solutions that make the search engines the way they are, as well as their current limitations and opportunities for further development.

According to Michael Mauldin [7], Chief Scientist at Lycos, a search engine is a program that performs the task of information retrieval according to some criteria, typically in the form of a search query. Due to the way that these programs are implemented, however, the term "search engine" has been generalized to include the construction of the information storage from which subsequent search operations are performed.

According to this definition, all major search engines perform two fundamental tasks: pre-compilation of a very large index of information (which may be at different levels of abstraction, such as individual Websites, Web pages, down to the individual words separated by white spaces); and subsequent retrieval of documents in response to a search query. The former entails an effective and efficient means of representing huge amounts of information while at the same time facilitates subsequent retrieval. The latter entails finding the closest match (often in a subjective sense) between the indexed documents and the search query. In addition, some search engines also attempt to gauge the implicit quality of a Web contents without making any reference to specific queries.

Among the earliest search engines were the Archie directory service [8], World Wide Web Wanderer [9], Aliweb (Archie–Like Indexing of the Web) [10], JumpStation [11] and the World Wide Web Worm (WWW Worm) [12], all of which were released between 1990 and 1993.

Archie was developed to fetch anonymous FTP (File Transfer Protocol) files to match filenames entered by the user. It therefore served as an index for FTP files. The World Wide Web Wanderer was actually a gathering program for the Web by checking Web servers on the Internet. It was designed to be used in conjunction with a retrieval program known as *Wandex*. As its name suggests, Aliweb uses an Archie-like mechanism to index the Web. More specifically, participating Web servers must first register with Aliweb. Each server in turn is required to construct an index of all pages on that server. Retrieval is then conducted on these pre-collected indices. JumpStation used a so-called *Web spider*[1] to collect title and header information from Web pages and used a simple exhaustive match to retrieve pages. The WWW Worm, on the other hand, indexed titles and URLs.

One distinguishing feature of the above search engines is that they simply produced a list of matching Web documents in response to a search query without any reference to the relevance of the retrieved documents. More sophisticated search engines such as the Repository–Based Software Engineering (RBSE) Spider [13], WebCrawler [14] and Lycos Spider [4] became available around 1994. Other popular search engines such as those from Yahoo! [2], Excite [3], and so on, soon became available. Although anecdotal evidence suggests that different search engines perform differently on different types of search queries (looking for people, news articles, etc.) [15], most of the popular search engines employ similar

[1] Also known as *crawler* (because it represents a program that "crawls" from one Web page to others via hyperlinks) or robot (which emphasizes its autonomous nature)

indexing and retrieval strategies. It is, therefore, worthwhile considering the underlying indexing and retrieval techniques involved.

2.3.1 Indexing

Most contemporary search engines employ the same fundamental operations to search and gather Web documents for processing and storage of information that characterize the documents. The aim is to produce an index database that minimizes storage requirements and facilitates subsequent retrieval. While many search engines create and maintain their databases using automatic indexing spiders, some (e.g. Aliweb and Yahoo!) build their databases (directories) manually. For example, Yahoo! defines a hierarchical structure (e.g. Home > Business and Economy > Shopping and Services > Jewellery > Watches or Home > Entertainment > Consumer Electronics > Audio > Audio Formats > Compact Disc (CD)). In directory-based systems, Web content providers typically submit their Web pages to be indexed to the search engine's editors manually. The editors then decide whether to include the submission, and the whole process can take as long as six months. Automatic indexing entails two operations: crawling (to search, gather and process documents), followed by storage of the index information in a database.

2.3.1.1 Crawling

In Step 1, the crawling operation begins with a non-empty set of Web pages arranged in some order. In Step 2, a starting point is selected to begin the search. Typically, the exact location of this initial point is not particularly important due to the highly connected nature of the Internet. Step 3 extracts distinguishing features (information like the header, URL, title, etc.) from the selected Web page under consideration. In Step 4, Web pages related to the selected Web page via hyperlinks are processed in the same fashion one-by-one until all such links are exhausted. Finally, in Step 5, this process is repeated until all the Web pages defined in Step 1 have been processed.

Various search engines differ from each other in the exact implementation of the above operations. For example, while some employ a queue to hold the set of Web pages to be processed in Step 1 above, others define a hierarchical structure. Other significant differences between search engines are the order of processing the Web pages in a set (Step 5) and determining what features to extract from the Web pages being processed in Step 3.

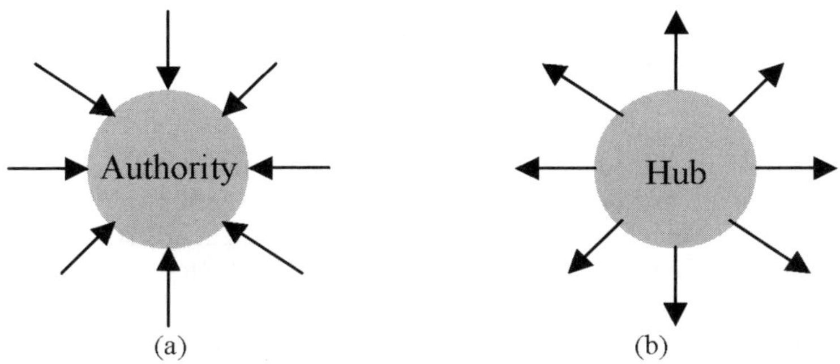

Figure 2.1 Illustration of Web pages that can be considered (a) an authority that is pointed to by many other Web pages and (b) a hub that references many other Web pages

In terms of choosing the order of processing in Step 5, popular techniques revolve around depth-first, breadth-first and so-called *best-first approaches*. The depth-first approach amounts to processing the most recently found page, followed by the next recently found pages, and so on. This approach has been known to cause severe overloading on servers and is now not commonly used [16]. It is also possible that important Web pages at the same level of the "seed" are missed as the user delves deeper down the chain of Web pages originated from the seed. The breadth-first approach favours the least recently found pages and is more suitable for small Websites with a relatively flat structure. So-called *best-first approaches* rely on some form of heuristics based on the concepts of hubs and authorities (see Figure 2.1) with the aim of identifying popular (and presumably of high quality) Web pages.

Indeed, there have recently been significant research activities in exploring the quality and applications of these hub/authority Web pages. In a more general sense, research into the relationships of Web pages via hyperlinks has shed light on the interesting problem of identifying the order of Web pages that should be traversed by a Web spider during indexing (in Step 5 above). Some of the most important research results on hyperlinks have been reported by a group working on the Clever project [17] and [18]. Other significant research in this area includes (but is not limited to) the Web Archeology project [19] and Google [1, 20].

In all these research activities, the ultimate purpose is to create an index such that subsequent retrieval will yield web pages that are relevant and of high quality in response to user queries. In this context, a relevant Web document is *not* necessarily also of high quality. For example, anyone interested in the subject of information theory will likely find Shannon's classic work [21] both relevant and of high quality. Subsequently written work based on this paper (e.g. students' summary of Shannon's work) is likely to be relevant, but less likely to be of the same quality. The study of hyperlinks in this context, therefore, aims to achieve the dual goal of relevance and quality.

Intuitively, a good hub provides a good survey on a subject matter (e.g. neural networks for multimedia processing) with many links to documents of high quality that provide more detailed information on specific topics (e.g. application of decision-based neural networks (DBNN) to image tagging). An authority is itself an important (if narrowly focused) page on a particular topic.

From the above discussion, it is apparent that hubs and authorities are somehow related. In fact, Kleinberg's Clever project group [18] have identified a strong interrelationship between authorities and hubs: a document that is linked to by many good hubs, and a good hub is a document that links to many authorities. Moreover, they have developed the hyperlink-induced topic search (HITS) algorithm for identifying authorities and hubs. Using their algorithm, good quality Web documents have been found to have high authority scores and vice versa, suggesting a correlation between the quality of a Web document and the authority score. This approach has been subsequently refined by the Clever group and others, for example, by introducing weights to the different links as in [22, 23].

Ongoing research in the analysis of the interrelationships between different Web pages via hyperlinks is underway. Further, some researchers are also attempting to analyse and model the structure of the Web. For example, some research groups (e.g. [18, 20]) are attempting to model the Web as a graph. These research activities will likely lead to better indexing algorithms in the near future.

Another major different between the various Web spiders exists in the different information that is extracted from each Web page being processed in Step 3. This is a tradeoff between storing too much information (that puts strains on storage) and too little information (so that indexing is not effective). Earlier Web spiders such as WWW Worm stored only Web page titles. Increasingly, due to the competitive nature of different search engines (as well as declining costs of storage and computing resources), contemporary spiders typically store entire documents for processing.

Once entire documents are stored, the spider performs textual processing typically by searching and counting the number of predefined keywords in the document. These keywords may be weighted according to their location within the document. For example, keywords found in the title, header and the top few lines of the document may be weighted more heavily than the same keywords found elsewhere.

The spiders also perform different statistical analysis on the results, often based on some undisclosed heuristics due to the commercial nature of these programs. In essence, the spiders attempt to extract some form of characteristics out of the documents processed.

Due to the unregulated nature of the Internet and commercial competitive pressure, occasionally, there are content providers who attempt to manipulate the results returned by search engines. Search engines that rely heavily on textual analysis of Web page contents are particularly vulnerable. For example, if a content provider deliberately inserts many occurrences of certain keywords (in non-displayed portions of a document), any algorithm that relies heavily on textual analysis would likely be influenced. The consideration of hyperlinks reduces this risk due to the following assumptions [20], which seem to agree with intuition:

- A hyperlink from a document *X* to another document *Y* suggests that the author of *X* endorses *Y* either fully or partially.
- If documents *X* and *Y* are connected by a hyperlink, then they might be on the same or similar topic.

On the other hand, using hyperlinks alone would not be fully reliable since one who could artificially inflate the "quality" of a page appear as an authority simply by creating many dummy pages that link to it. Thus, the consideration of hyperlinks provides additional reference in determining the nature of a Web document. In this context, combining Web page analysis based on keywords and link information leads to yet another way of determining the relative importance of different Web documents in relation to a given topic.

For example, it is possible to obtain a weighted average of results from three strategies in assigning an importance factor to each document. The three strategies are link-based (e.g. hub/authority score), similarity-based (e.g. pattern matching between keywords and selected topic), and a combination of the two. Coming up with good heuristics and techniques to complement this kind of approach remains an open research problem.

Finally, it is important to minimize the storage requirements of the final results. In particular, the extracted indices should occupy only a small fraction of the storage needed for the original documents.

2.3.1.2 Storage

The actual means by which different search engines store the extracted information is treated with the same degree of commercial sensitivity as the processing described above. However, the databases all share some common characteristics. The engineering decisions involved in the creation and maintenance of each database are heavily influenced by both the dynamic nature of the Internet and the large number of Web documents that are available. Another requirement is that the information must be stored in such a way that facilitates effective retrieval. Finally, an additional requirement of all search engine implementation is cost effectiveness in all its operations

Therefore, the characteristics of a database for storage of the extracted information on Web pages are:

- *Trouble-free maintenance* — these databases must be designed to facilitate (frequent) updating of records. Such updating activities include the modification of existing documents, the addition of new documents and the deletion of documents that no longer exist.
- *Space efficient* — since these databases typically store information on tens or even hundreds of millions of Web documents, they should be designed to minimize usage of disc space.
- *Effective retrieval* — the records stored in the database must be well organized and be easily accessible during retrieval.

2.3.2 Retrieval

In essence, retrieval of indexed Web documents amounts to sifting through the information in the database to find the best match between the user query and the indexed documents. This observation leads to two important considerations in the design of an effective retrieval mechanism:

- Computation of a matching score that can be used to determine how close an indexed document matches the search query prompted by the user. Clearly, this is useful not just for finding relevant documents but also for ranking the retrieved documents.
- Formulation of user queries that is both user-friendly and amenable to the computation of the matching score.

It is worth noting that the response to a user search query is not necessarily a list of results derived from the above considerations. This "distortion" is attributed to simple financial considerations. Since purveyors of search engines typically generate revenues from advertisements, many search engines display so-called sponsored links either as the top-ranked retrieved results or otherwise feature prominently together with other retrieved results (e.g. in a sidebar next to the main list of results).

2.3.2.1 Scoring and Ranking

The mechanism used in scoring and ranking database records in response to a user search query mirrors that used in the processing of Web documents during the Indexing stage. The major differences are the following:

- The finite set of records to be analysed come from the database, rather than from a queue (or some other subset) of all Web documents on the Internet.
- Statistical analysis performed on the database records is based on the user query, rather than the predefined set of keywords.

Based on the above observations, the scoring and ranking stage therefore entails analysis based on some or all of the following:

- Counting the number of query terms that appear in each database record.
- Counting the number of occurrences of each of the query terms in each record.
- Determining the distance between the occurrences of the query terms in each record. A suitable distance measure must therefore be pre-selected. Typically, this can be in terms of the number of words between a pair of query terms.
- Introducing weights to the different query terms found in each record based on the location of their occurrences. For example, query terms found in the title or header may be assigned heavier weights than the same terms found elsewhere.
- Determining the similarities between the query terms and words or phrases that might have similar or equivalent meanings in each record. For example, "3G" might be identical to "third generation", as are the terms "cellular phone" and "mobile phone".

It should be noted that a query term might contain a phrase of multiple words in general. For example, in the case of a Boolean search, a permissible query may contain the phrase "(information theory or communications) and average codeword length and Shannon". In addition, the quality of a query also affects the ability of a search engine to find the information that the user really finds useful. A recent study of over 200,000 users has found that while search topics have changed, there is little change in user search behaviour [24]. Consequently, users should also be encouraged to adapt to the new environment so as to be able to formulate effective search queries. On the other hand, Web designers and content providers should also address this problem.

2.3.2.2 Query Formulation

Regular users of search engines will be familiar with the form of queries that are likely to yield useful results, as opposed to those that do not. In general, generic terms (e.g. "the"), terms that can have different meanings in different contexts (e.g. "net") and popular names (e.g. "John Smith") need further qualification to narrow down the list of possible matches. For example, a popular name should be entered as a search query along with another factor such as a place of work, to yield the desired result.

Regardless of the search engine in use, search queries are typically entered phrases of words, often with Boolean operators such as "not", "and", "or", parenthesis, and so on. Some implementations also have special operators, such as the pipe "|" operator that facilitates progressively narrow searches within a selected topic.

Once a query has been entered, it must be compared with the database records. Earlier search engines such as WWW Worm performed an exhaustive search to compare the query against each record. This resulted in a list of results ordered not according to relevance, but according to the locations of the records stored in the database.

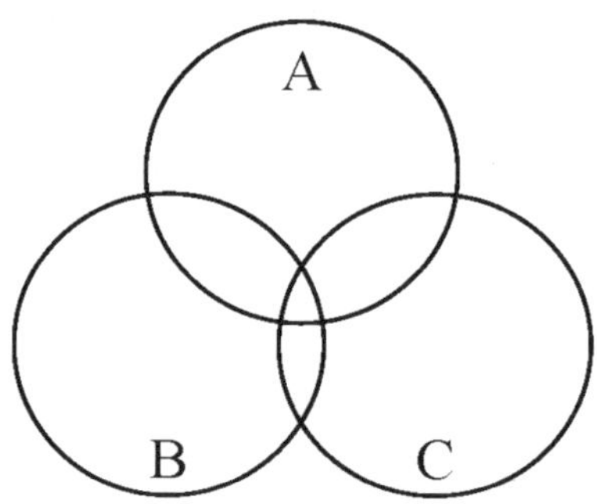

Figure 2.2 Illustration of inverted file indexing. Analysis is performed on the intersection of the sets of records that contain the individual query terms A, B and C, respectively. Clearly, the search space is greatly reduced compared to any exhaustive search of the entire database, or all members of sets A, B and C

Contemporary search engines typically employ a so-called *inverted file indexing* scheme, where an inverted file is defined as a list of all occurrences of words in the text database. For each unique word in the database, the search engine maintains a list of documents containing that word. Sometimes it also maintains a list of all the positions where the word occurs. In this way, retrieval analysis and score matching can be performed on those sets of records that represent the intersection of those sets that contain the individual query terms as illustrated in Figure 2.2. This is clearly more compute-efficient than an exhaustive source of all database records, or the union of all the sets concerned.

2.3.2.3 Similarity Measures

During the indexing and retrieval phases, it is important to be able to quantify similarities between keywords and documents, and between database entries and search queries, respectively. For example, suppose that during the retrieval phase each index document is represented by a vector d that characterizes the document, and the query is represented by a vector q, preferably of the same dimension as d. Then, it is possible to calculate the Euclidean distances $D_j(d_j,q)$ between q and all n candidate documents in a collection (where n is finite e.g. as in the intersection shown in Figure 2.2) using the equation

$$D_j(d_j,q) = \frac{\sum_{i=1}^{n}(wd_i \times wq_i)}{\sqrt{\sum_{i=1}^{n}(wd_i)^2 \times \sum_{i=1}^{n}(wq_i)^2}} \quad \text{for all } j = 1 \text{ to } n \tag{2.1}$$

where wd_i and wq_i are the weight of the ith element in the d and q, respectively.

Since this equation measures the Euclidean distance between the document vector and the query vector, the document $j = j_{min}$ having the minimum value of $D_j(d_j,q)$ in a collection is given the highest ranking. This is based on the observation that the documents with a lower Euclidean distance to the query will be semantically and conceptually closer to it [25]. While other similarity measures may also be applied, the authors have found the Euclidean distance measure to be simple and effective in actual implementation.

2.3.2.4 Query Independent Ranking

This is an attempt to gauge the intrinsic quality of a Web document without making any reference to the input query. The result may also be combined (possibly in the form of a weighted average) with a result obtained using some other query-based algorithm.

One of the best-known query-independent ranking algorithms is the PageRank algorithm [26], which is used by Google. The intended purpose of this algorithm is to separate low quality Web documents from high quality ones. The PageRank value $R(X)$ of a document X is given by

$$R(X) = \frac{\varepsilon}{n} + (1-\varepsilon).\sum_{(X,Y)\in C}\frac{R(Y)}{N_h(Y)} \tag{2.2}$$

where ε = constant selected from the range 0.1 and 0.2 inclusive
n = number of Web pages in a collection
C = collection that includes pages X and Y
$N_h(Y)$ = number of hyperlinks on page Y (also known as the outdegree of Y in the literature)

The PageRank value is therefore computed recursively for each Web page in a collection based on the number of hyperlinks found on the referencing page. In particular, PageRank of a page X is computed by assigning weights to each hyperlink to X proportional to the quality of the page Y that contains the hyperlink. The quality of the referencing page Y is in turn computed by applying PageRank recursively.

2.3.3 Meta Search Engines

From our above discussions, it is clear that although contemporary search engines have a lot in common at a conceptual level, there is significant scope for variations in actual implementations that account for the apparent differences in their behaviours. More often than not, results generated by these search engines are quite different in response to the same search query. Meta search engines, such as Net7 [27] and MetaCrawler [28] have been developed to combine the strengths of the various search engines so that they complement each other.

The basic idea behind meta search engines is to enable a parallel search using multiple search engines in response to a user query. In the case of Net7, results from the different search engines are simply listed in different windows without performing any subsequent comparison or ranking on the results obtained from the individual search engines. However, it does provide very specific operators such as "narrow" to search for very specific topics. MetaCrawler, on the other hand, performs ranking on the results from the various search engines and presents a single ranked list of results to the user.

At any rate, there are common features that a meta search engine should possess (as summarized in Figure 2.3):

- A unified "one-stop" user- interface
- A preprocessing operation that takes the user search query as an input and outputs a number of search engine-specific query formats that correspond to the different formats adopted by the various search engines used. For example, AltaVista prefers searching for phrases by enclosing the phrase in double quotes, e.g. "entropy coding".
- As searching times are likely to vary with different search engines, it is important to synchronize the search operations and/or provide a user-definable time limit.

- A postprocessing operation that takes the result lists from the various search engines and collates all these results in a consistent manner. Duplicate results should be discarded. Ideally, the postprocessing operation should also rank the collated list of results.

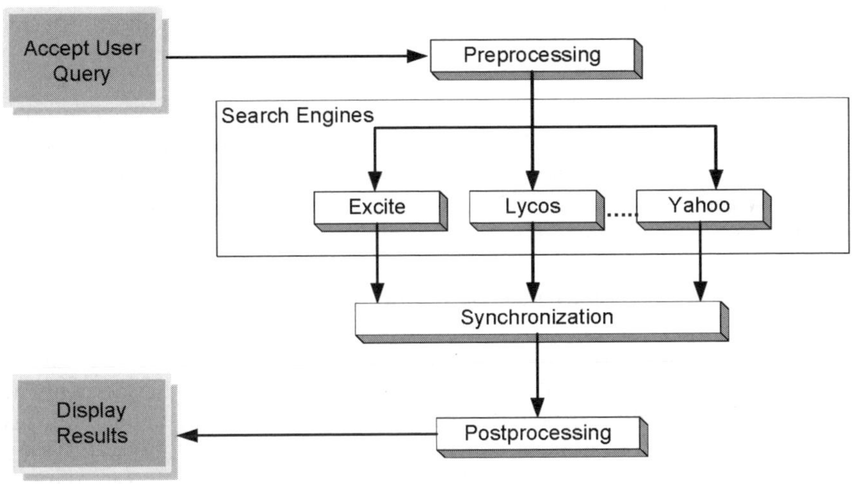

Figure 2.3 Basic operations of a meta search engine

In addition, the meta search engine should be adaptable and scalable to handle the ever-changing nature of the Internet.

2.3.4 Non-Technical Limitations of Search Engine

Amidst all the technological advances in the development- and the popularity- of search engines, they still have many limitations, some of which are not related to technical issues. Quite apart from the quality and number of results returned in response to a query, one of the major limitations of all search engines is their coverage of the entire Web. A study performed by Introna and Nissenbaum [29] found that none of the search engines indexed more than 16% of all Web pages, and only 42% were indexed by the search engines combined. This is hardly the forum for global and fair exchange of information that many people (including the pioneers and architects of the Internet) would like the Internet to be.

This observation begs the important question: "what causes a Web page to be included or excluded by the search engines?" A generalization of this is the relative prominence of different Web pages: Web content providers (especially those of a commercial nature) want not only to have their pages indexed by popular search engines, but also feature prominently in any returned lists of results in response to search queries.

While the increasing number of non-English/non-textual Web contents pose new challenges in multilingual and multimedia processing, which may cause the exclusion of some Web pages with non-English/unrecognized data formats, Introna and Nissenbaum argue that the factors determining the prominence of Web pages are economical as much as they are technical. For example, strategic alliances between search engines and service providers mean that the returned results can often be biased.

In summary, economical- even political- forces can exert a strong influence on the results presented by search engines, just as much as technical issues do. Consequently, many specific applications, such as the discovery of scientific publications (discussed in Section 2.6), require dedicated solutions that may or may not utilize generic search engines.

2.4 PERSONALIZED MONITORING SERVICES

The capability of search engines that lets users find useful information on the Web empowers both content providers and end-users. This has certainly helped popularize the Internet, particularly the Web, as a source of up to date information. Taking this a logical step further, many users would like to be able to keep track of changes in the Web contents of individual interests. This is precisely the purpose of personalized monitoring services (PMS).

2.4.1 Current Web Monitoring Systems

While using search engines to access information may be considered a pull operation, monitoring is analogous to a push operation through which previously requested information is pushed to the user [30, 31] when the information becomes available or updated. For example, a user may be interested in the score of a football match as soon as this is known.

Many specialized monitoring systems [32] and [33] have been developed to monitor specific Web information that users are interested in. However, the monitoring domains tend to be rather limited and specific. The most common monitoring domains include research publications, books, music CDs, stock quotations, online auctions and online news. These systems are developed specifically for the targeted domains. They provide the necessary user interface to let users specify the required information. These systems then monitor the information from the predefined sources based on user specification. However, it is very troublesome to employ several specialized monitoring systems for users who wish to monitor information from several domains.

A number of generic Web monitoring systems such as NetMind [34], TracerLock [35], WatzNew [36] and WebCQ [37], have been developed to provide monitoring without domain-specific limitations. These systems also allow users to monitor Web page components such as links, images and text, or any portions of a Web page. Figure 2.4 illustrates a classification of current personalized information monitoring systems, both domain-specific and generic.

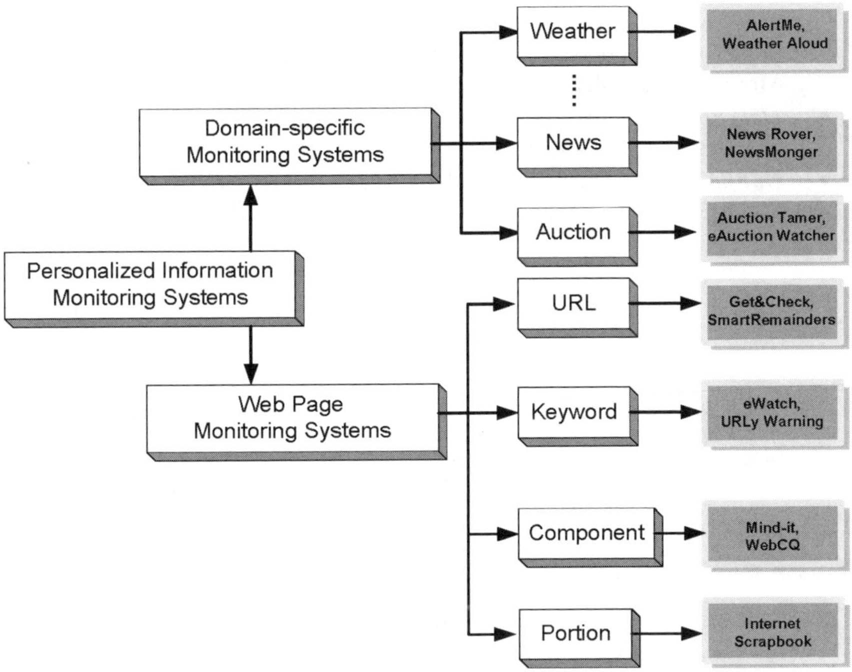

Figure 2.4 Classification of personalized information monitoring systems

However, most Web monitoring systems available are only able to track changes on the entire Web page, or certain components (attributes) such as images and links. This is very inflexible, as users are unable to freely specify whatever they want to monitor from the targeted Web page. Some systems also support Text Minding [34], which makes use of a "copy-and-paste" operation to allow users to specify any portions of information on a Web page to monitor. However, such an operation is not user-friendly, and in many cases, the user does not know how to specify the desired information to monitor.

2.4.2 An Alternative Web Monitoring Model

Figure 2.5 shows a Web monitoring model that consists of four major processes: specification, extraction, detection and notification:

- *Specification* — Using the block heading-tree approach (described below), this process lets users specify their interest on a Web page to monitor. Users can select the blocks or items in a block to monitor. Users can also monitor images, links or keywords within a block. The user-specified information is passed to the extraction process.

- *Extraction* — This process extracts the block heading-tree of the Web page for use in the specification process. After the Specification process, this process stores the user-specified information and related data into the Monitoring Database.
- *Detection* — This process is activated periodically to check for possible changes on the monitored Web pages.
- *Notification* — Once changes have been detected, this process is activated to notify the users concerned about these changes via e-mail.

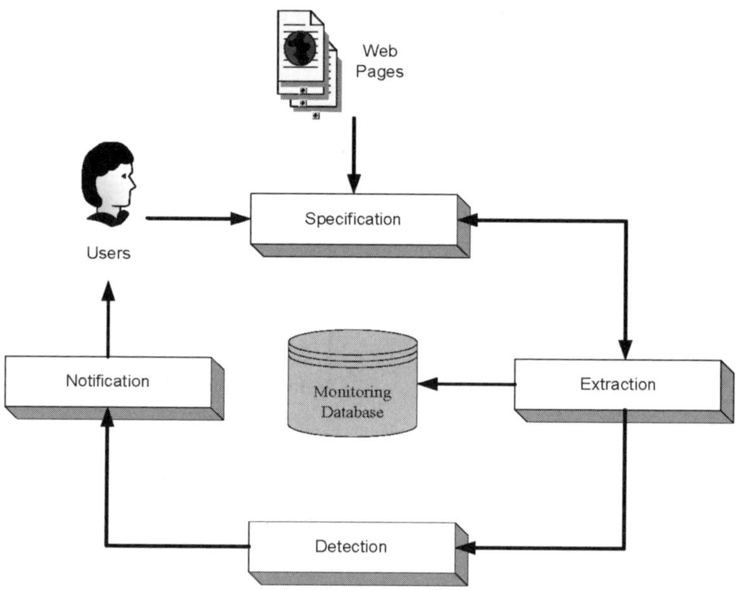

Figure 2.5 Overview of a Web monitoring model

2.4.2.1 Block Heading-Tree

The formulation of block heading-tree is based on the consideration of Web pages as a hierarchical structure consisting of blocks of formatted and unformatted data as illustrated in Figure 2.6. A Web page can be viewed as a hierarchy of three levels: Web page level, block level and item level. The first level is called the *Web page level.* It simply represents the entire web page content in the document that may contain a hierarchy of blocks. The second level is called *Block Level.* Each block can be classified as structured or unstructured. Structure blocks are tables and lists, whereas unstructured blocks are data blocks located between structured blocks, or those that are outside structured blocks. Unstructured blocks may include data items such as text (paragraphs or formatted paragraphs), links and images. Structured blocks may also be nested. The third level is called the *Item Level*. All the data located inside structured blocks will be in the form of block items such as table items and list items. The data located inside an unstructured block will be treated as a single block item for the unstructured block. Other elements such as links, images and text can also be included inside a block item.

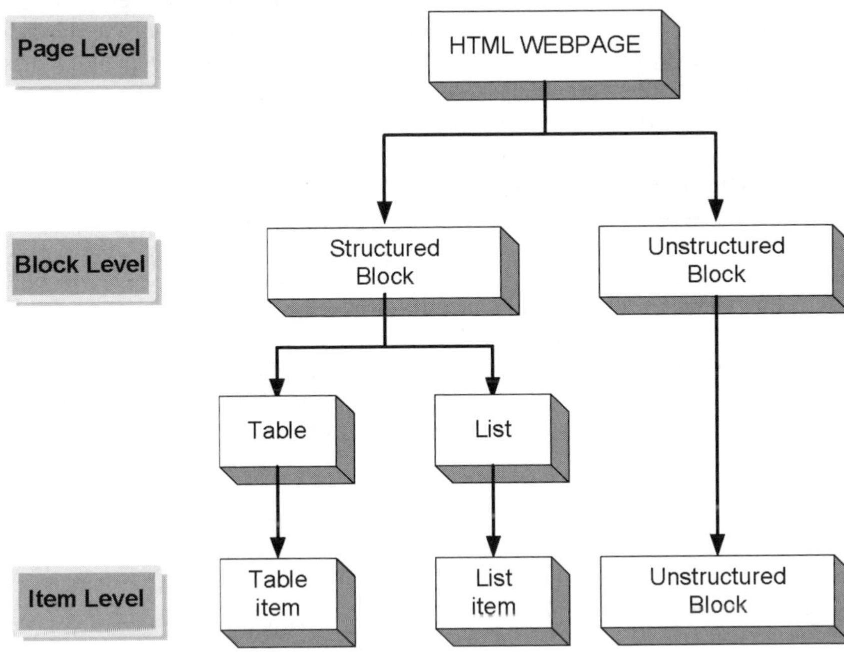

Figure 2.6 Three-tier representation of a Web page

The purpose here is to generate an efficient data structure that preserves the hierarchical structure in order to facilitate user-specified block-by-block monitoring. In fact, a user can easily specify the monitoring of a complete block or individual items within a block. The block heading-tree is displayed graphically as a user-friendly interface for the specification of blocks/items to monitor.

2.4.2.2 Specification

This process serves as an interface between the user and the system with the use of the block heading-trees generated by the extraction process. Figure 2.7 illustrates the specification process. First, a Web document is extracted and transformed into a block heading-tree by the extraction process. The user may then select the block to monitor all the data items located inside the block, or select an individual item to monitor. Furthermore, the user can specify particular types of information, such as images, links and keywords within a block. In keyword specification, the user can specify certain keywords to further refine the information that the user wishes to monitor from the selected block. Overall, three types of selection are permissible: *block selection, item selection* and *component selection*. These are discussed below with the aid of Figure 2.7.

- *Block selection* — *Today's Forecast* is the block that the user selects. The entire block is also highlighted on the Web page in response to this selection. The system stores the monitored information and keeps track of any changes in this particular block.
- *Item selection* — In the block *Today's Forecas*t, there are several items such as Temperature, Wind Speed, Air Pollution, and so on. For example, if temperature is the only item of interest, then the user can select only *Temperature* in the block *Today's Forecast* to monitor.
- *Component selection* — In the block *Today's Forecas*t, the user may have interest in the change of weather conditions, such as sunny, cloudy, and so on. Then, the user can monitor image change in this block. Currently, the image in this block is cloudy. If there is no change in the weather tomorrow (i.e. the cloudy image remains), then no change will be detected. The link option works similar to the image option. The keyword option is useful when users want to detect any changes on specific topics or concepts.

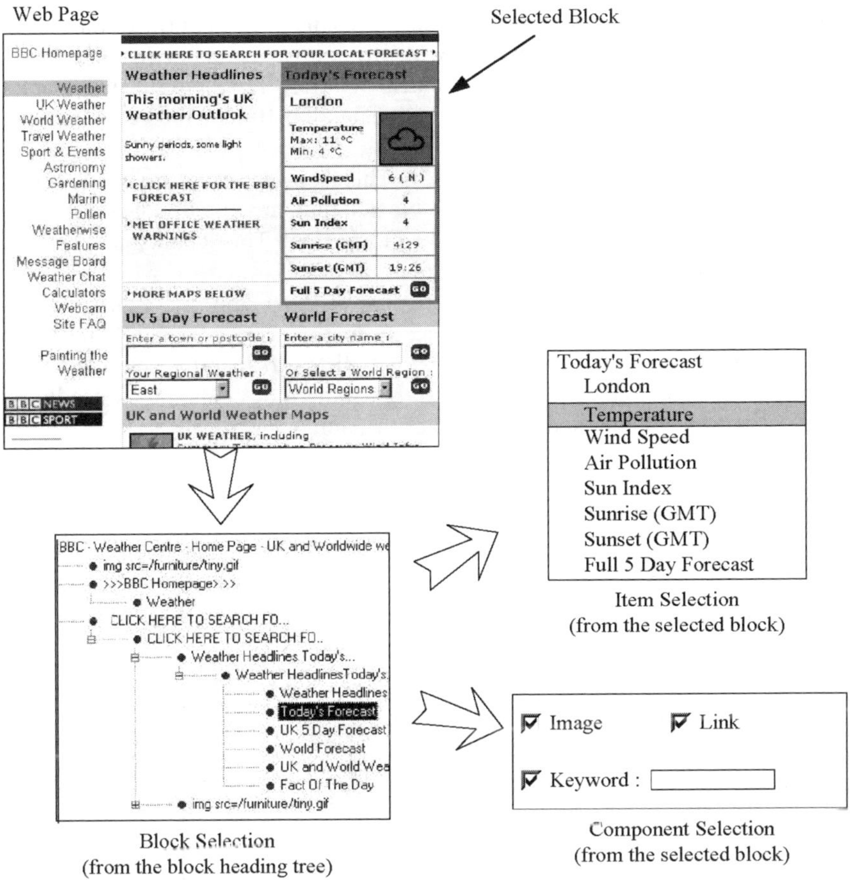

Figure 2.7 Illustration of the specification process

2.4.2.3 Extraction

The main objectives of the extraction process are to extract the block heading-tree of each monitored Web page for use in the specification process, and then to store the monitored information into the monitoring database for future change detection. Figure 2.8 shows the heading-tree generation process that comprises the following four steps: Parsing Web Pages, Building Block Trees, Locating Block Items and Identifying Block Headings:

- Step 1 — *Parsing Web Pages* — This step parses the text data and tags information on a Web page to form a text string and a tag list structure, respectively.

- Step 2 — *Building Block Trees* — This step identifies unstructured blocks, and structured blocks with structures such as tables and lists. For nested block structures, hierarchical relationships between parent blocks and child blocks are also identified. The block tree is the basic framework for the block heading-tree.
- Step 3 — *Locating Block Items* — This step identifies the block items information. In addition, nested blocks that are mainly used for layout purposes are removed from the block tree. This results in a modified block tree with block item information.
- Step 4 — *Identifying Block Headings* — This step uses heuristics to extract the heading for each block from the source code. If the heading information cannot be identified, it will try to identify suitable text data from the source code as the heading for the block. The block heading-tree structure is generated at the end of the step.

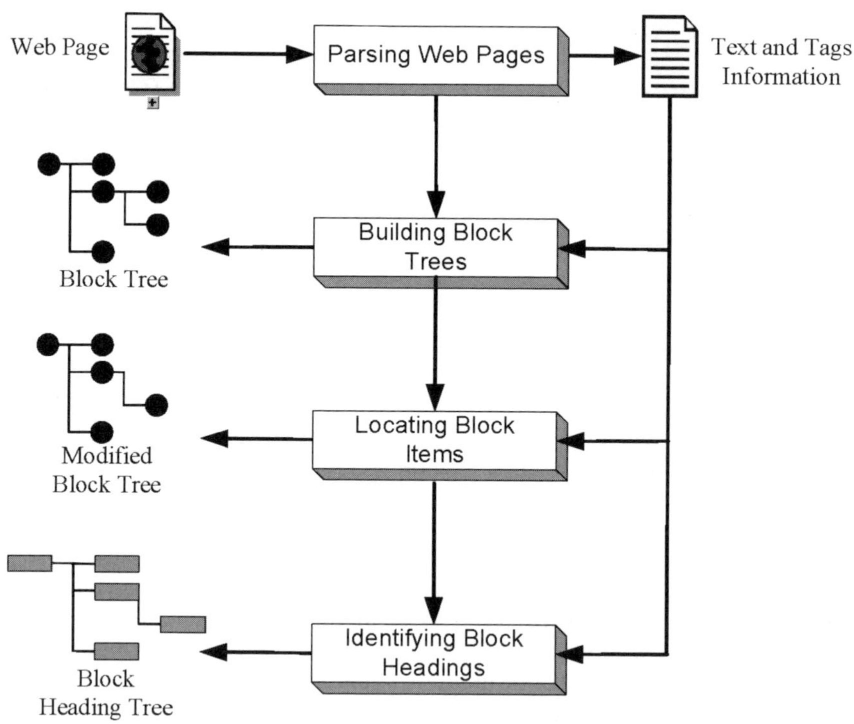

Figure 2.8 Illustration of the extraction process

Besides generating the block heading-tree for the specification process, the extraction process also stores user-specified information from the specification process into the monitoring database. Four groups of data are stored in the database: general information, monitored block information, user-option information, and heading-tree information:

- *General information* — It includes the URL address of the Web page and the date of extraction. The URL is used to retrieve the Web page during the detection process. If a change is detected, the date of extraction will be used in the notification report to indicate the date the user has specified this monitoring action.
- *Monitored block information* — It includes the block number, block heading, block items, block text data, and block tags. Among them, the block number and block heading are received from the specification process. Block items, block text data, and block tags are extracted from the block structure of the heading-tree.
- *User-option information* — It includes additional (optional) information such as item, image, link and keyword, all of which are specified by the user using the specification process.
- *Heading-tree information* — It contains the block heading-tree data. It only contains the block heading and child blocks. Information on block items is not stored.

2.4.2.4 Detection

The detection process is executed periodically. Entries in the monitoring database are retrieved and processed. All detected changes are sent to the notification process. As shown in Figure 2.9, this process consists of five sub-processes: Generating Updated Heading-Tree, Loading Original Heading-Tree, Comparison, Sending Change Message, and Updating Monitoring Database.

- *Generating Updated Heading–Tree* — This uses the URL field in the monitoring record as input. It downloads the Web page, parses the HTML document, and generates the new block heading–tree. The mechanism is the same as heading–tree generation in the Extraction process. The new heading–tree is called *updated heading–tree*.
- *Loading Original Heading–Tree* — This generates the heading–tree directly from the heading–tree field of the monitoring record. The result is called *original heading–tree*.
- *Comparison* — This is the main function of the detection process. It uses the monitored block information and compares the updated and original heading–trees. Various types of changes, such as block change, item change, image change, and so on, can be detected.
- *Sending Change Message* — When changes are detected, a message that summarizes the changes is sent to the notification process.
- *Updating Monitoring Database* — This updates the monitoring record in the database as appropriate.

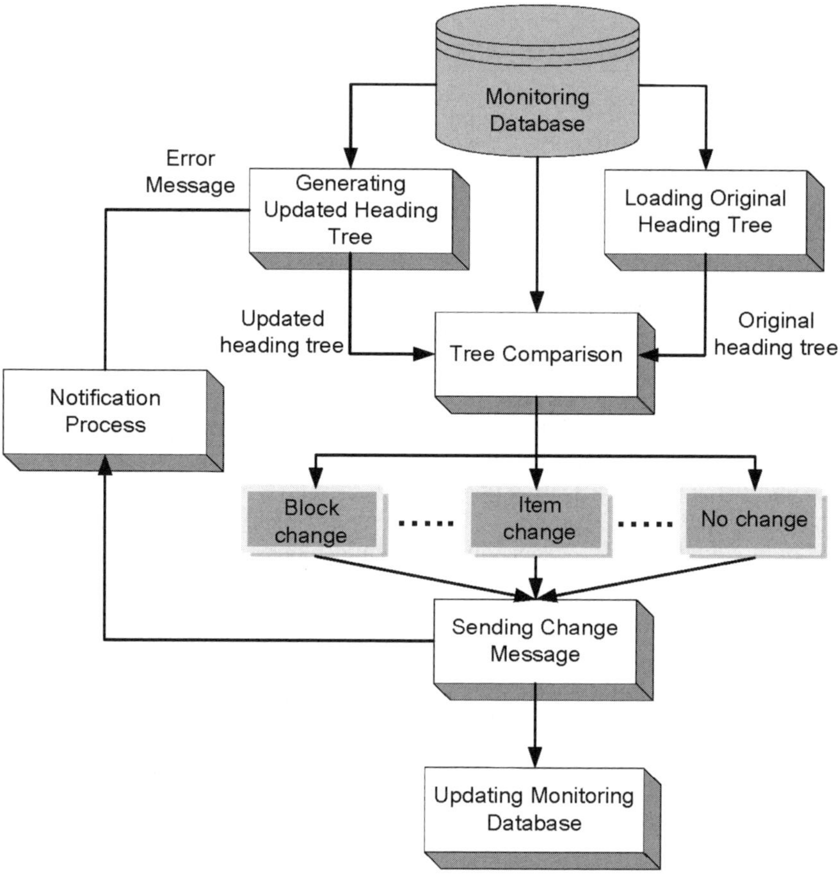

Figure 2.9 The Detection process

2.4.2.5 Notification

As the final stage in the Web monitoring model, this process provides mechanisms for the generation of change summary reports to the users. The change summary report in the Web monitoring model consists of three groups of information as follows:

- *User monitored block information* — It tells users what monitored information has changed. It includes the block heading and the URL.
- *Changes detected* — It tells users all the changes that have been detected. The types of changes that may be detected include: block/item change, block/item removal, image change, link change and keyword change.
- *Link to the changed contents* — Contents of the old block and the updated block can be displayed via the link.

Notification e-mail

Subject: Change Detected from WIM
Date: Thu, 8 Aug 2002 11:20:04 +0800
From: wim@ntu.edu.sg This is Spam | Add to Address Book
To: bfbingfeng@yahoo.com

Dear bfbingfeng@yahoo.com,

From your specifications on

URL: http://www.bbc.co.uk/weather

Monitored Block: Today's Forecast

The following changes have been detected

* Item Change
* Image Change

Check the changes from your personal WIM or by clicking the URL link above.

Thank you for using WIM, your personalized Web Information Monitoring system!

Figure 2.10 An automatic e-mail notification generation by WIM

In this process, the notification service is provided by server-initiated push delivery. Once changes are detected, they are delivered to the client immediately. Finally, the change summary report is sent to the users via e-mail as shown in Figure 2.10.

2.4.3 The Web Information Monitoring System (WIM)

The Web monitoring model described in Section 2.4.2 has been implemented as the Web information monitoring system (WIM). Figure 2.11 shows the WIM's client-server architecture.

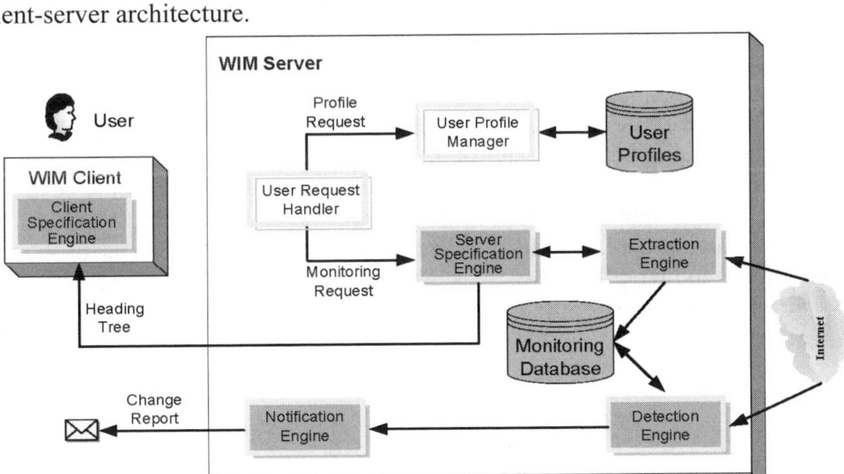

Figure 2.11 An overview of the WIM's client-server architecture

Figure 2.12 shows the WIM client interface developed using Visual Basic. A user can click on the icon 'M' to start monitoring. The news window is used to inform the user of any new updates. The user can then enter a URL of interest. The WIM client will pass the URL to the *User Request Handler* in the WIM server, which is implemented in Java, to generate the block heading-tree.

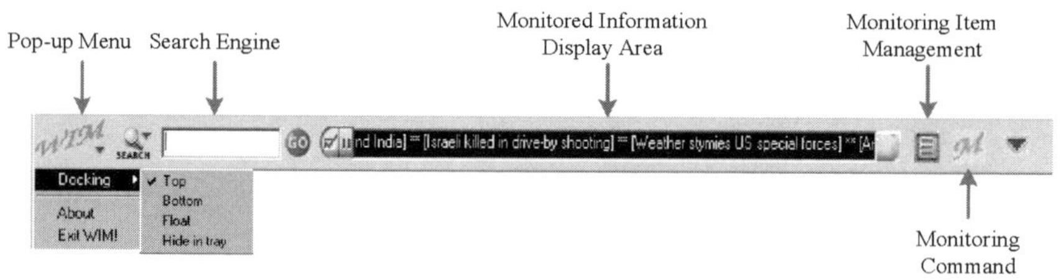

Figure 2.12 WIM client interface

The *Monitoring Manager* receives the monitoring request and calls the *Extraction Engine* to generate the heading–tree, and then passes the heading–tree to the *WIM client* for display to the user. The user can click on a tree node to select a block to monitor, and inform the *Monitoring Manager* to save the monitoring request into the *Monitoring Database*.

A Timer is set to be active every 24 hours to periodically invoke the *Detection Engine* to evaluate all the monitoring requests stored in the database. It calls the *Extraction Engine* to download and extract the updated Web pages and retrieves the stored Web page from the database. The *Detection Engine* sends change messages, if found, to the *Notification Engine*, which generates a change report and sends to the user. If the *User Request Handler* receives a *profile request* from the WIM client, such as user login, or request for new account creation, the *User Profile Manager* handles it by interacting with the *User Profiles* database.

Figure 2.13 shows an example of client-side display of block heading–tree. The left hand side of the screen shows the actual block heading–tree. The user selects the tree node *Today's Forecast*, which is a structured block, in the heading–tree. Then, on the right hand side of the screen, the corresponding block on the Web page is highlighted. The block heading–tree, therefore, helps users to view the content of the block they have selected. If this is the block of interest, the user can perform the corresponding monitoring operation.

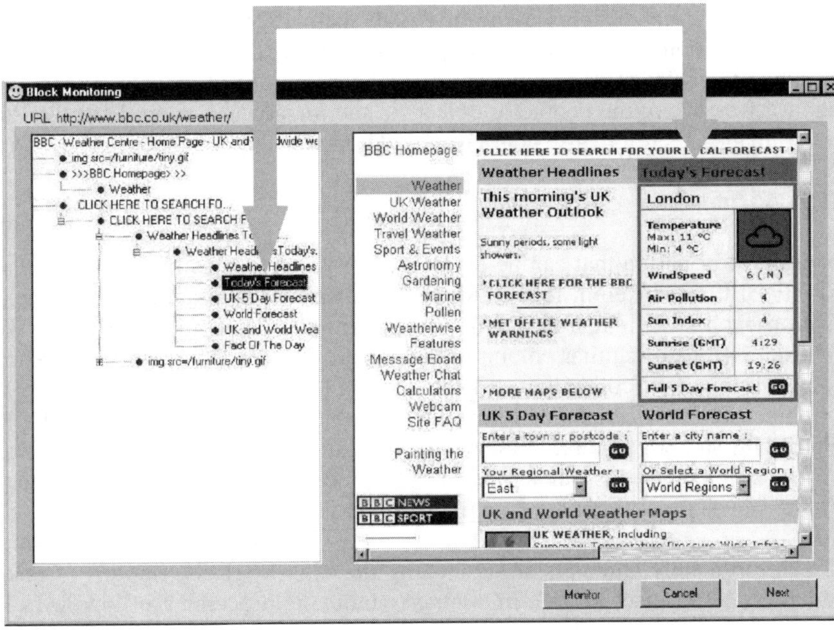

Figure 2.13 Example of client-side display of a block heading-tree

2.5 STORAGE AND RETRIEVAL OF VISUAL DATA

The discussion so far has focused primarily on textual information. With the advent of broadband Internet access, however, it becomes feasible for users to obtain transfer visual data files, which are typically large compared to textual data, via the Internet. Indeed, there is an increasing demand for visual data such as images and video clips. In response, many Internet content providers also increase both the variety and the number of visual data files.

The growth of visual data on the Web creates new problems in the development of effective tools for end-users to search and retrieve the ever-increasing collections of such files. This is because solutions that work well for textual data are not generally good for visual data. In fact, a good solution for effective storage and retrieval of visual data calls for advances in storage, image/video processing, indexing, retrieval and transmission. This section presents a survey of current state of the art solutions to this end.

2.5.1 Images

Images are characterized by three fundamental attributes: colour, texture and shape. These attributes are collectively known as *visual features*. In addition, some images also have non-visual features associated with them. Non-visual features typically include information about an image in question, such as the date and time the image was captured/updated/stored, as well as textual descriptions of the image such as the location it was taken, the people or objects in the image, and so on. Effective solutions for images, therefore, use both visual and non-visual cues to aid analysis.

2.5.1.1 Visual Cues

The visual information that can be extracted from an image can be considered as a hierarchy of three different levels. At the lowest level, individual pixels carry information such as colour and intensity. Collections of neighbouring pixels form edges and regions of similar colour and intensity. Texture information can also be extracted from these collections. At the second level, these collections of pixels can be interpreted as semantically meaningful objects (e.g. a red car, a dark piece of cloth with smooth texture, etc.). At the highest level, the juxtaposition between the various objects forms human-intelligible concepts that describe the image, for example, "an eagle in flight" or "a child playing with a dog".

Unfortunately, attempts at automating the high-level descriptions of images have largely been unsuccessful. In contrast, automatic processing of low-level features has been more successful. In fact, current image indexing and retrieval systems rely on semi-automatic methods that involve some form of manual processing in the introduction of high-level descriptions often in the form of textual information that accompany an image.

Low/mid-level visual features: These are colour/greyscale, texture and shape. There are currently three commonly used colour models, namely, the Commission Internationale d'Eclairage (CIE) chart, red-green-blue (RGB) cube, hue-saturation-intensity (HSI) or hue-saturation-value (HSV) space. The RGB cube is the most relevant here as it is mostly used for the display of digital images. A description of these three colour models can be found in Appendix A. Greyscale information is treated in a similar manner as RGB representation, only simpler because each pixel has one associated greyscale value instead of the three values for the R, G and B components. Greyscale images are therefore treated as a special (simplified) case of colour images in this context.

Texture information is much more difficult to measure than colour information. Intuitively, texture can be described as "smooth" or "rough", but describing it quantitatively is much harder. In image processing, texture is considered as complex visual patterns composed of entities that have characteristic intensities, colours, slopes, sizes, and so on. There are two major approaches: statistical analysis and structural analysis. Statistical analysis measures the variation of intensity in a portion (window) of the image. The purpose is to derive some form of statistical signature that characterizes the window. Neighbouring windows that have similar or identical statistical signatures might be merged together to form a larger region of homogeneity in terms of texture.

Shape information is even harder to obtain and quantify as compared to texture because real-life objects typically have irregular and non-rigid shapes. In addition, occlusion and incomplete objects (e.g. an image showing half a car) further complicate this task. Extracting shape information from an image requires the segmentation of conceptually meaningful objects. Although object segmentation is a task that human beings take for granted, attempts at developing automated systems have largely been unsuccessful and only semi-automatic systems can perform reasonably well. This is because some form of semantic understanding is generally required. The application of artificial intelligence has been found useful to some degree in this context.

2.5.1.2 Non-Visual Cues

Image-specific textual information can be used in conjunction with visual information in the characterization of images. Textual information that is often used as non-visual cues includes date/time/location of capture, date/time of storage (or most recent update), resolution, file type, compression method, as well as annotations. However, since annotations and query terms are entered by persons, it is important to impose some rules to ensure consistency in the description of the same/similar images.

2.5.1.3 Application of Artificial Intelligence (AI)

Artificial neural networks (NN) have been found useful in the development of content-based image retrieval systems. For example, a system that utilizes decision-based neural networks (DBNN) has been found useful. Developed by Yu and Wolf [38], the system performs offline tagging of images using a DBNN. Moreover, it uses a multi-resolution approach to reduce the search space when looking for a feature in an image. Each image is classified offline into a series of predefined subjects using colour and texture information with DBNN. Figure 2.14 gives an overview of the system's visual search methodology.

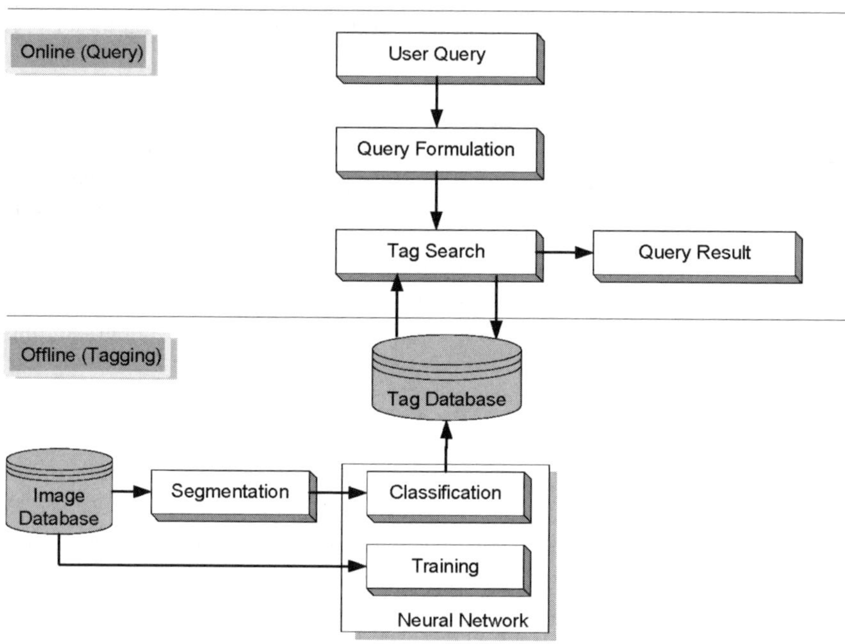

Figure 2.14 Visual search methodology

Since queries are answered by searching a tag database and images are not manipulated online, the system is capable of fast online performance. In contrast to other content-based systems (that use similarity-based retrieval), this system supports subject-based retrieval, allowing users to perform a semantic search. The 4-step tagging process, which is illustrated in Figures 2.15 and 2.16, is as follows:

- Divide each image into 25 equal-sized blocks, each of which contains an arbitrary number of objects.
- Use colour information for initial classification. Each block is classified into one of following families in the HSI colour space: black, grey, white, red, green, blue, yellow, magenta and cyan.
- Use texture information to refine the classification results. Each block may be classified into one of the following categories: sky, foliage, flesh-tone, wood, and so on.
- Save tag in the tag database.

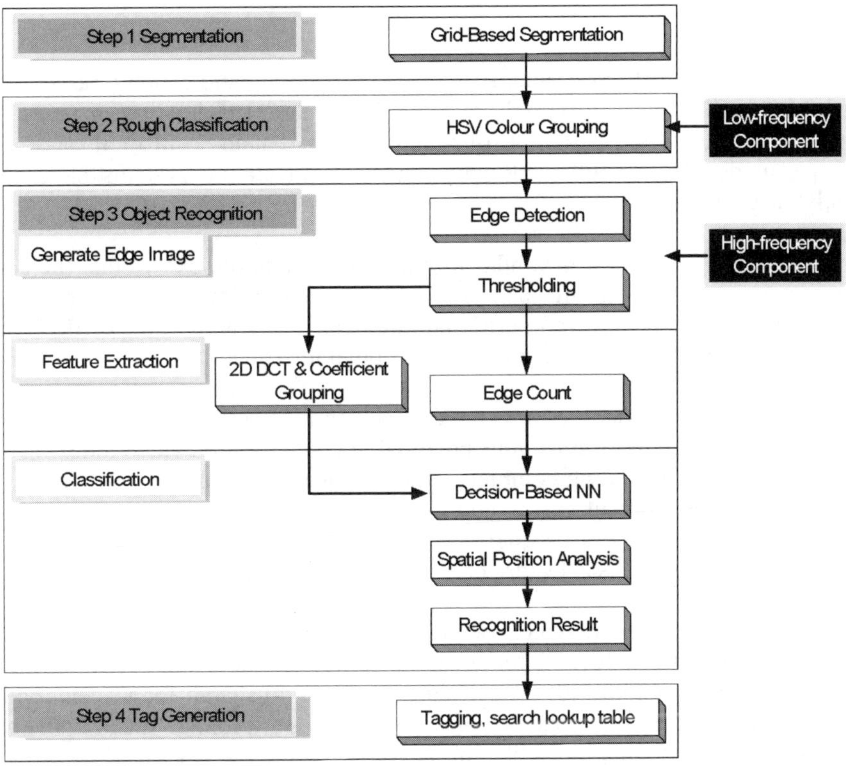

Figure 2.15 The tagging procedure

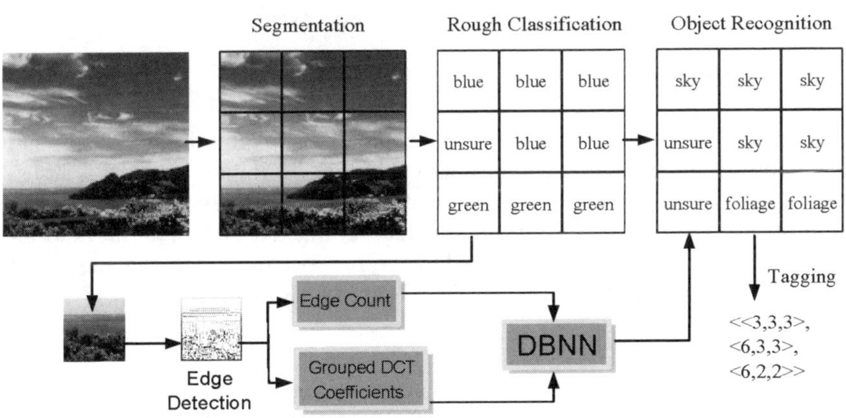

Figure 2.16 Illustration of the tagging procedure

2.5.2 Videos

With the advent of MPEG-4, interactive video processing and content-based video retrieval have become increasingly feasible due to the consideration of video objects. However, the segmentation of video objects remains a difficult task, and attempts at developing fully automated video segmentation systems have been largely unsuccessful.

Video segmentation is similar to image segmentation, except that the additional inter-frame motion information provides more opportunities and challenges for analysing video data. Again, artificial intelligence can help in the difficult task of video segmentation and higher-level conceptual interpretation.

The ability of machines to extract semantically meaningful objects or events out of video footage has tremendous potential in content-based video indexing and retrieval, not just for finding complete files, but also for finding specific frames in a video sequence. Content-based (and more specifically semantic) video retrieval allows a viewer to use search queries such as "find Nelson Mandela's inaugural speech" or "find Prof. XX's lecture on interactive multimedia broadcasting".

Currently, however, typical search queries for video clips are by means of the specification of low-level features (e.g. colour, brightness and texture information) [39] or by example [40]. For instance, a query to search for a scenic video may be "Find frames with 70% sky blue on top and 30% foliage green at the bottom" or "find a scene that looks like this example".

Clearly, this is inadequate, and machine understanding of video footage can bring automation to the indexing and retrieval processes.

2.5.2.1 Application of AI

An important practical application of AI to video retrieval is in face-based indexing and browsing. In many video applications, it is important (but tedious) to browse through large amounts of video material to find a relevant clip. Researchers have proposed the use of NN for video database indexed by human faces to facilitate search. For example, Lin *et al.* [41] have proposed a 3-step scheme based on face detection and recognition using DBNN as follows (see Figure 2.17 for an illustration):

- *Step 1* — Segment the video sequence by applying scene change detection that gives an indication of the first and last frames of a particular shot. Each segment is considered a "story unit" of the sequence.
- *Step 2* — Use a probabilistic DBNN face detector to identify segments with high probability of containing human faces.

- *Step 3* — Representative frames from Step 2 with high probabilities are annotated and used as indexes for browsing.

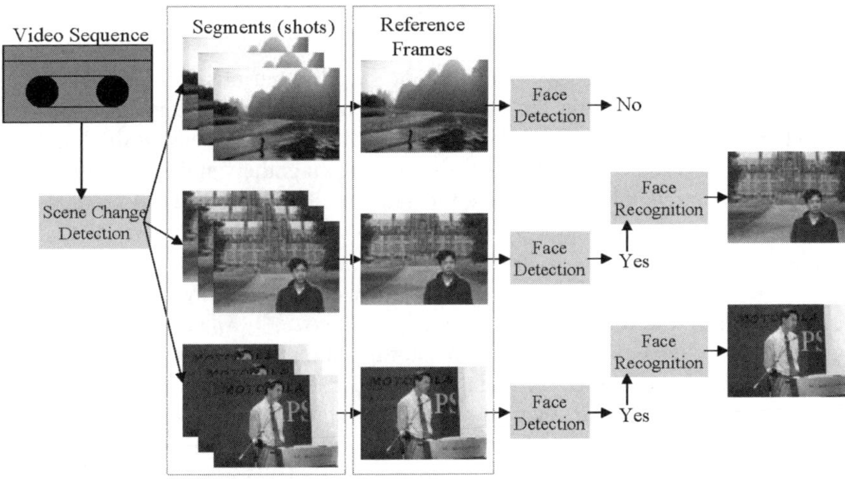

Figure 2.17 Illustration of a face-based indexing and browsing that relies on AI

A scene change can be detected using any of the following approaches:

- *Histogram comparison* — if the histograms of two consecutive frames differ by a predetermined threshold, scene change is considered to have occurred between the two frames.
- *Motion vector comparison* — if an abrupt change in motion vectors is detected between consecutive frames, a scene change is likely.
- *Comparison of transform coefficients* — if the video data has been compressed (e.g. in MPEG [Moving Picture Experts Group] format), then corresponding DCT (discrete cosine transform) or DWT (discrete wavelet transform) coefficients can be compared between consecutive frames.
- *Analysis of non-visual cues* — occasionally, textual information is also available to aid scene change detection. For example, an optical character recognition (OCR) system could be used to read captions for further analysis. This would also aid semantic understanding of the scenes themselves.

Methods for face detection and face recognition are less well established as those for scene change detection. In general, these two operations are accomplished using some form of AI. A suitable NN (a probabilistic DBNN in the above example) is trained using a sufficiently large sample to generate a knowledge base. If the NN is well trained, then the knowledge base represents a generalization of knowledge on detecting/recognizing different faces under different conditions.

This knowledge base is then used in the recall phase to detect/recognize faces from previously unseen video sequences.

2.6 Case Study: Discovery/Monitoring of Web Publications

Various systems have been developed and deployed in various domains for solving specific problems when using the Web as an information repository. For example, the LawBot system [42] has been developed for the specific needs of the legal profession, as well as other interested parties. The system gathers and organizes statutes and case histories available on the Web to facilitate legal search both by legal professionals and laypersons.

Meeting the needs of the biomedical professionals requires yet another set of problem-specific solutions. For example, Lovell *et al.* [43] have described techniques for using the Web as a tool for the storage and remote retrieval of biomedical signal data. The chief problem associated with biomedical data is that there is a lack of wildly accepted standard, even though some standards do exist (e.g. American Society for Testing and Materials (ASTM) 1467 for biomedical signals and the Institute of Electrical and Electronics Engineers (IEEE) 1073 medical information bus standard) [44]. This is because manufacturers of medical measurement instruments tend to use proprietary data representation formats.

Discovery and monitoring of special interest Web documents, such as scientific publications, also require problem-specific solutions. This section presents a case study of the discovery and monitoring of scientific publications on the Web in order to illustrate the challenges involved in making the Internet work for special needs. Generic search engines such as Google are good as a starting point to search for specific items. Manual browsing through the numerous results can be tedious and frustrating. A number of systems have been developed to cater for the specific needs of the scientific community. The purpose is to empower researchers with the ability to search for relevant scientific publications available on the Web. Taking this a step further, monitoring systems that concentrate on scientific publications on the Web assist researchers to keep track of latest developments in their fields of interest.

2.6.1 Discovery of Web Scientific Publications

Many scientific publications are now available online over the Web or stored in the form of Digital Libraries. However, the information available on the Web tends not to be well organized, making the search of relevant publications difficult and time consuming. Generic search engines such as Yahoo!, Lycos and Excite have mostly proved ineffective for searching scholarly publications accurately.

Autonomous Citation Indexing Agents such as CiteSeer [45] have been developed to search computer science-related publications on the Web. They extract citation information from the publications and store it into a citation database, which is frequently updated using the Citation Indexing Agent. The citation database contains rich publications information that can be mined for subsequent retrieval in response to user queries.

2.6.1.1 CiteSeer

Using autonomous agent technology, CiteSeer locates, parses and indexes scientific publications found on the Web and generates a citation database. It supports two types of keyword searches on citations and indexed publications. When searching for citations, all citations that match the given query, as well as the context of source papers containing the citations, are retrieved. The results are ordered according to the number of times each paper is cited.

When searching the full text of indexed publications, CiteSeer returns the header for matching publications along with the context of the publication where the keywords occur. Users can order the publications according to the number of citations, or by publication date. CiteSeer can also display related publications. The relatedness is calculated using several algorithms. A Term Frequency x Inverse Document Frequency (TFIDF) (explained in Section 2.6.1.4) scheme is used to locate publications with similar words. Distance comparison of publication headers is used to find similar headers. Common Citation x Inverse Document Frequency (CCIDF) (explained in Section 2.6.1.4) is used to find publications with similar citations.

Pubsearch is another system that is based on co-citation analysis to relate publications in a citation database. PubSearch differs from CiteSeer in that it also supports document cluster searches apart from the traditional cited reference searches. The related publications are grouped into clusters based on common keywords found in their citations. As such, users can retrieve all the related publications even though some publications may not contain the exact keywords supplied by the user.

2.6.1.2 PubSearch's Citation Database

Scientific publications typically include some references for the reader to probe further. A citation index contains the references that a paper cites, linking the source paper to the cited papers. Citation indices can be used to identify existing research fields or newly emerging areas, analyse research trends, find out the scholarly impact, and avoid duplication of previously reported work.

A citation database stores these citation indices. It contains all the cited references published with the articles. These cited references reveal how the source paper is linked to prior relevant research because the citing and cited references have a strong link through semantics. Therefore, citation indices can be used to facilitate the searching and management of information. Some commercial citation index databases such as those provided by the Institute for Scientific Information (ISI) [46] are available on the Web.

Figure 2.18 shows the structure of the Citation Database, which consists of two tables, SOURCE and CITATION. The SOURCE table stores information of the source papers while the CITATION table stores all the citations extracted from the source papers. Most attributes of these two tables are identical, for example, paper title, author names, journal name, journal volume, journal issue, pages and year of publications. Full text access of linked articles is possible through the URL link stored as one of the attributes.

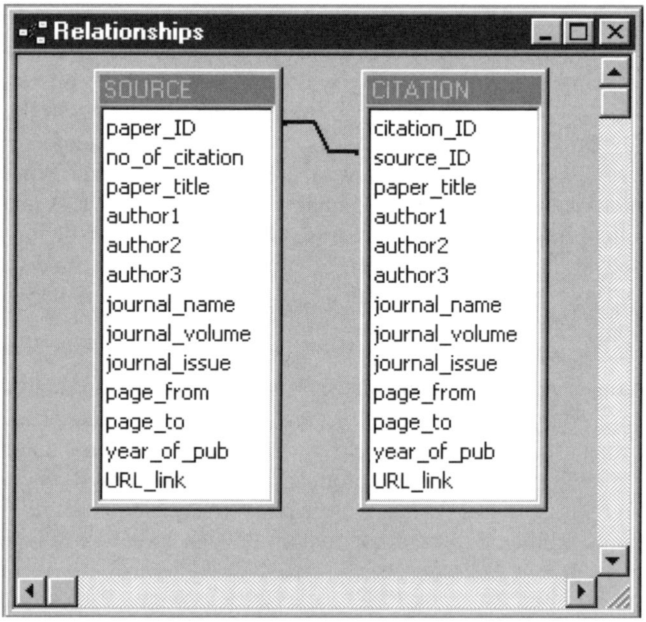

Figure 2.18 Citation database structure

The primary keys in these two tables are "paper_ID" of the SOURCE table and "citation_ID" of the CITATION table, respectively. "no_of_citation" of the SOURCE table indicates the number of references contained in the source paper. "source_ID" of the CITATION table links to the "paper_ID" of the SOURCE table to identify the source paper that cites the particular publication stored in the CITATION table. If two different source papers cite a publication, the publication will be stored in the CITATION table with two different citation_IDs.

An experimental test Citation Database has been developed [47] by downloading publications from 1987 to 1997 in the Information Retrieval (IR) field of the Social Science Citation Index from ISI. A total of 1,466 IR-related papers were selected from 367 journals with 44,836 citations.

2.6.1.3 The PubSearch System

The PubSearch system [47] has been developed to effectively mine a citation base for the retrieval of scientific publications on the Web. Figure 2.19 shows the PubSearch system architecture. The term "publications repository" in the figure is used to underline the fact that the same retrieval framework could be applied to documents stored in a digital library or some other databases.

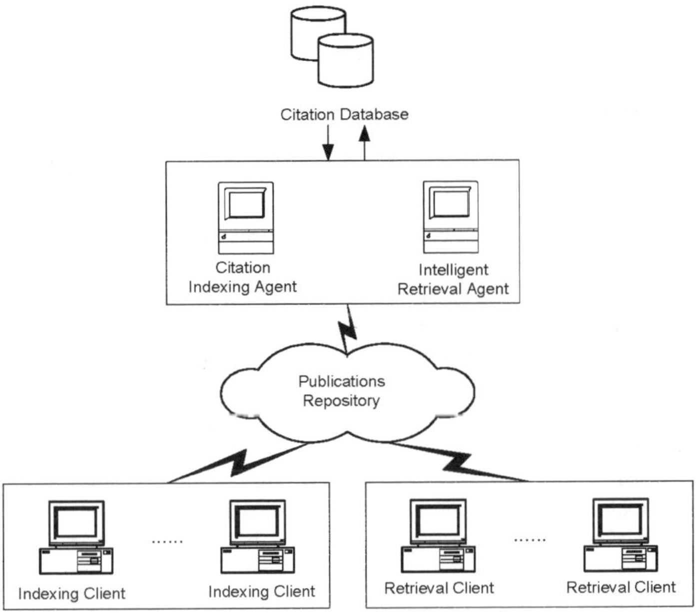

Figure 2.19 Overview of the PubSearch system

The two major components in PubSearch are the Citation Indexing Agent and Intelligent Retrieval Agent. The Citation Indexing Agent employs two approaches to find scientific publications. The first is similar to CiteSeer, which uses search engines to locate Websites containing publication keywords. The other is to allow users to specify publication Websites through the Indexing Clients. The Citation Indexing Agent then downloads the publications from the Websites and converts them from PDF (Portable Document Format) or PostScript format into text data using the *pstotext* tool [48].

The bibliographic section of Web publications is identified through keywords such as "Bibliography" or "References". Citation information is then extracted from the bibliographic section, and stored in the Citation Database. The Intelligent Retrieval Agent mines the citation database to identify the hidden relationships and explores the useful knowledge to improve the efficiency and effectiveness of publication retrieval.

In addition, the system provides a number of Indexing Clients and Retrieval Clients. These clients serve as an interface between the user and the system. Through an Indexing Client, users can specify the Websites to be included, as well as the frequency of visits to these sites by the Citation Indexing Agent. A Retrieval Client provides the necessary user interface for query entry that will be passed to the Intelligent Retrieval Agent for further processing.

2.6.1.4 Application of AI to PubSearch

Citation information can be used to judge the relevance of publications in response to a search query because authors cite articles that are related. The measure TFIDF (term frequency x inverse document frequency) is commonly used to determine the relatedness of different documents [25]. TFIDF is computed as follows. Each component of a document vector is calculated as the product of Term Frequency (TF) (i.e. number of times word w_i occurs in a document) and Inverse Document Frequency (IDF) (i.e. log [D/DF(w_i)] — where D is the number of documents and document frequency DF(w_i) is the number of documents in which w_i occurs at least once). Using this method, the documents can be classified into different groups according to the distance (which measures the similarity or relatedness) between them.

In PubSearch, keywords are extracted from each document's citations. In the Citation Database, no full text contents of cited articles are available. The keywords are extracted solely from the titles of all citations. Each extracted keyword forms an element of a document vector. If D denotes the document vector, then each keyword will be denoted by d_i where i is between one and N, and N is the total number of distinct keywords. For each document, 20 most frequently occurred keywords are extracted from its citations. The use of 20 keywords, which gave the optimum result, was determined experimentally. The TFIDF method can then be adopted to represent the document vector.

After solving the document representation and relatedness problem, a neural network (in this case a self-organizing map SOM [49]) is used to categorize documents in the citation database. This entails two processes: SOM training and retrieval. The training process mines the citation database to generate cluster information (NN knowledge base), grouping related documents into clusters; and the retrieval process retrieves and ranks the publications according to user queries through the Retrieval Client.

2.6.1.5 Retrieval of Scientific Publications Using PubSearch

Figure 2.20 shows the cluster map in response to the query "relationship between recall and precision" using the SOM NN. This example shows that semantic clusters are spread across several units. The best-matched cluster "95" is highlighted, which is displayed together with its neighbours. Thus, users can explore documents residing in these neighbouring clusters in addition to the best-matched cluster.

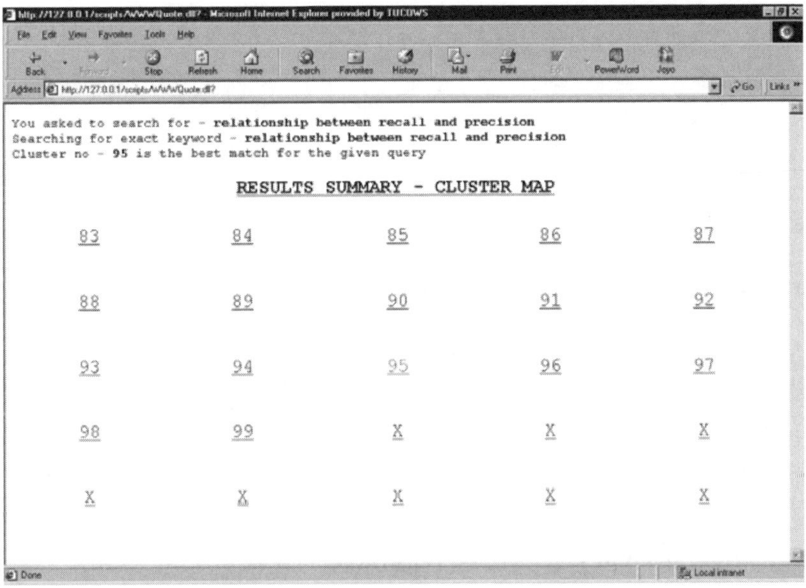

Figure 2.20 Summary of results presented in the form of a cluster map.

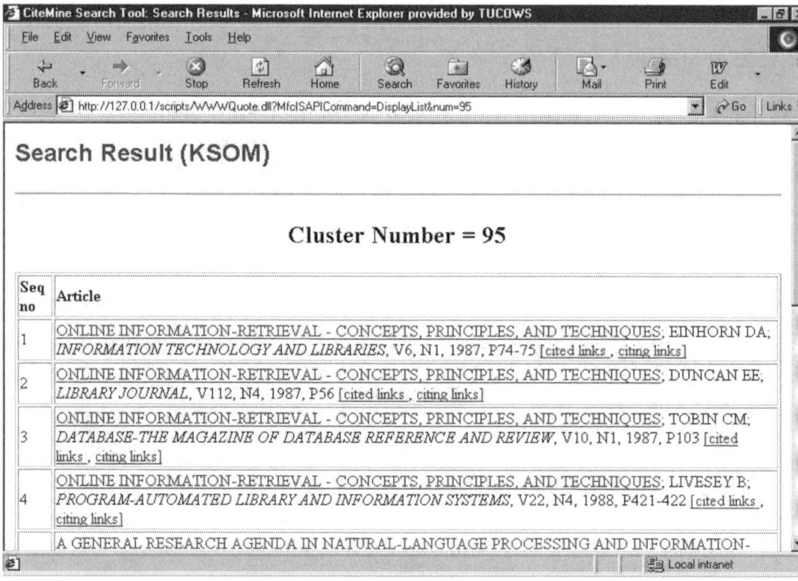

Figure 2.21 Ranked list of the retrieved publications in cluster number 95

By selecting any of the cluster numbers in the cluster map, documents from that particular cluster are listed and ranked according to the least Euclidean distance defined in equation 2.1. This is illustrated in Figure 2.21. The paper titles are underlined to allow the user to get the full text content of the paper via the URL links. There are also "citing" and "cited" links provided, which allow the user to go deeper into the citing or cited documents of that particular publication.

2.6.2 Monitoring of Scientific Publications

Research institutions and individual researchers increasingly choose to put up their publications on their Websites to facilitate the exchange of ideas with other researchers. These publications are usually listed in an index Web page. Generally, researchers are aware of certain research institutions or researchers who are renowned in a particular research field.

Since index pages are usually updated either periodically or whenever a new research paper is published, these are frequently visited and tracked for updates by other researchers in the field. Although Web browsers can be used to access the index pages, researchers still need to spend much time accessing the targeted Web pages manually. In addition, they sometimes have to browse through whole Web pages in order to locate the updated information. Such manual monitoring tasks are tedious and time consuming.

As discussed in Section 2.6.1, generic Internet search engines are not very effective in locating scientific publications. Similarly, commercially available monitoring systems are also not suitable for scientific publications. A dedicated monitoring system, known as *PubWatcher*, has been developed for tracking scientific publications from user–specified Web sites or pages [50]. PubWatcher differs from other systems in that it is not a digital library like CiteSeer, but a personalized monitoring service dedicated to Web scientific publications. PubWatcher allows users to define Websites or pages of their interest to explore. It then performs automatic retrieval and monitoring of relevant publications.

In PubWatcher, an effective publication extraction technique [51] has been developed to extract publication information from the publication index pages of the monitored Web sites. Although publications listed in the index pages are often displayed in formats similar to those of the bibliographic sections of scientific publications, no standard has been adopted for citing Web publications. As a result, each institution or research group has its own format and convention for displaying publication data. In addition, publication information may be displayed separately or mixed with other kinds of information. The different formats and organizations of publication index pages pose a challenging problem for monitoring systems to extract publication information from the index pages correctly.

2.6.2.1 Analysis of Web Publication Index Pages

Most publication index pages are HTML-formatted pages that list research publications including journal articles, books, technical reports and conference papers. There are two major types of index pages: dedicated index pages and mixed index pages. Dedicated index pages provide publications information only. On the other hand, Home pages set up by individual researchers are often mixed pages since they usually include other kinds of information such as the author's resume. In the publication listing section, the index page usually contains three major components: publication blocks, publication items and publication attributes, as follows:

- *Publication Blocks* — A publication block refers to the portion of a publication index page that contains all the publications. The block often contains a header and/or an introduction, followed by the list of publications and papers. An index page may contain multiple publication blocks.
- *Publication Items* — Each publication block usually contains multiple publication items, which are used to list the research publications or articles according to a certain format. If the publication has a hyperlink to an online version, this publication item is called an *online publication*.
- *Publication Attributes* — Each publication item contains attributes of the publication, such as the title, author(s), publication date, publication

name and series and pagination information. Online publication items also contain the document hyperlinks.

The most common citation standards for scientific literature are American Psychological Association (APA), Modern Languages Association (MLA) and American Medical Association (AMA). Although publications listed in the index pages stylistically resemble the reference section of a print article, some elements may be omitted or altered. For example, the URL address of an online publication may be explicitly listed or hidden in the hyperlink. As such, index page analysis is needed to extract publication information from the index pages [51].

2.6.2.2 Monitoring of Web Publication Index Pages

Figure 2.22 shows the Web publication index page monitoring process that consists of three major sub-processes: Web Page Extraction, Index Page Identification and Publication Extraction.

- *Web page extraction* — This process extracts all the Web pages in the same site from the provided URL by exploring the hyperlinks that connect to other Web pages.
- *Index page identification* — This process determines if a Web page is a publication index page based on keyword analysis and heuristics.
- *Publication extraction* — For each index page, this process extracts publication information of all the listed publications. The results are stored in a database.

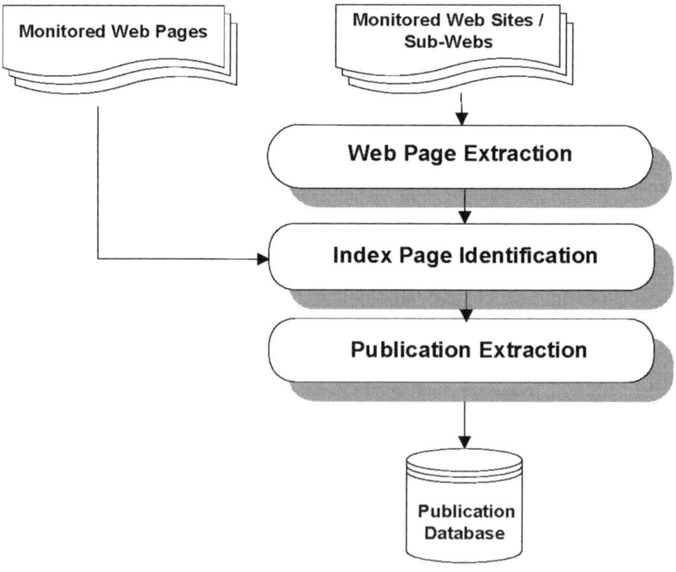

Figure 2.22 Web publication index page monitoring process

2.6.2.3 PubWatcher System Implementation

Figure 2.23 shows the architecture of the PubWatcher system, which is based on client-server architecture. The client serves as a Web-based interface for users to interact with the system. Several databases are also maintained to store different information for the operations of the system. The server contains the following major components:

- *User Management* — It maintains users' account information and monitors their system usage.
- *Meta Search Engine* — Multiple search engines are employed to search for interesting Web sites or pages for monitoring.
- *Monitoring Specification* — It allows users to specify interesting Web sites and pages for monitoring.
- *Monitoring and Tracking* — It monitors the user-specified Web sites/pages, and checks for any updates according to the frequency specified by the users.
- *Information Delivery* — It delivers new publications to the users' personal folders and notifies the users about the updates via e-mail.

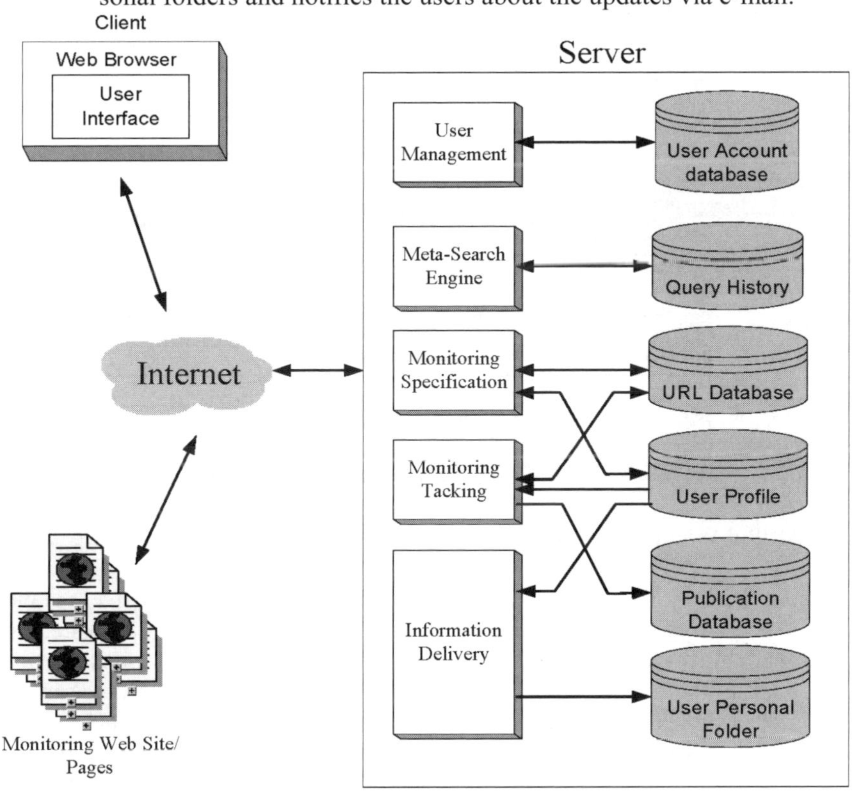

Figure 2.23 Overview of PubWatcher's system architecture

2.6.2.4 PubWatcher Retrieval

Each PubWatcher user is assigned a personal folder that stores all the information of their interested publications. Based on the monitoring results, the extracted publications that match the user's specifications on keywords, author names, publication name and/or publication date are saved into the user's personal folder. An e-mail is also sent to inform the user about the new updates. The e-mail message contains the number of new publications found, and the link to access the new updates via PubWatcher.

The retrieval client supports three types of retrieval: New Publications, All Publications and Search Publications. New Publications Retrieval displays a list of newly found publications. All Publications Retrieval supports the retrieval of all the publications stored in the user's personal folder. Search Publications Retrieval provides an interface for users to specify a search query, which will then retrieve all the publications that satisfy the search query.

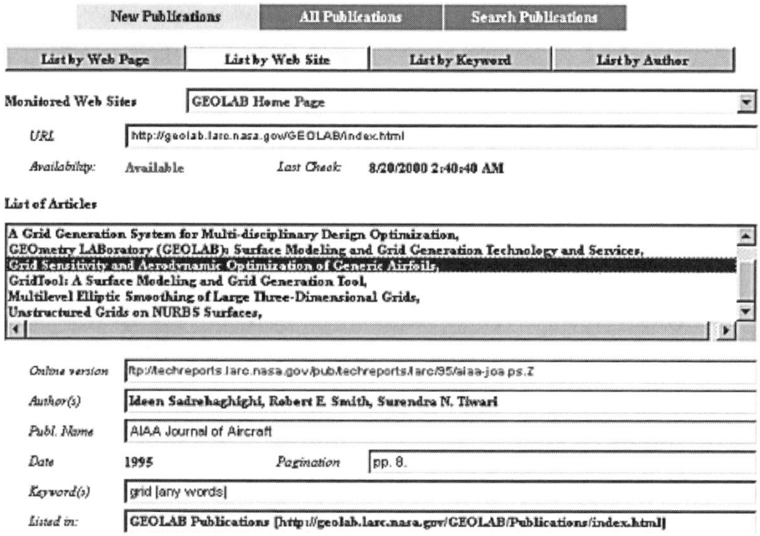

Figure 2.24 PubWatcher's retrieval interface

Figure 2.24 shows an example of New Publications Retrieval. The results are grouped according to the monitored Websites. For each monitored Website, the URL, the availability status and the last check date of the Web site are listed. Publications found in the monitored Website are also displayed. When the user scrolls down the publication list, attributes of the corresponding publication such as the URL of the online version, author, publication name, publication date, pagination information and other related information such as the keywords and the page in which the publication is listed, are displayed accordingly.

2.7 FURTHER ADVANCEMENTS

This section highlights technological advancements that are likely to bring benefits to the Internet community for the dissemination and retrieval of information. For example, semantic Web is an initiative to inject both semantic meanings and uniformity to Website development efforts that will hopefully make information retrieval easier. On the other hand, advances in natural language processing are likely to improve machine analysis of user search queries. Although most Web contents are in English, there is an ever-increasing amount of non-English Web contents being made available. Multilingual processing of Web contents is therefore needed to facilitate indexing and retrieval of Web contents in different languages.

In the areas of multimedia data retrieval, further research focuses on machine understanding of high-level descriptions of visual data to facilitate indexing and retrieval. This includes human-oriented descriptions, such as the extraction of facial expressions and gestures to aid high-level descriptions. In addition, speech processing technology will increasingly be applied to Internet search engines. Finally, the Internet (more specifically the Web) will certainly be used as an information repository in more and more domains, especially when the Internet is also considered a communication medium. Some of the most significant applications will be discussed in subsequent chapters.

2.7.1 Semantic Web

Developers of effective Web indexing and retrieval typically have to assume that their solutions must operate in an environment where documents are not necessarily well organized or presented. This is the nature of the Web as it stands, as opposed to organized collections of documents such as purpose-built digital libraries, in which documents are well organized to facilitate retrieval. This represents a shift of responsibility in ensuring that relevant information is accessible from search engine developers to content providers. What if the same paradigm is adopted by Web content providers in the wider context? Why not make the Web contents accessible by design?

The semantic Web initiative [52] is being developed by the W3C organization and other interested parties to facilitate human–computer interaction. The Semantic Web is an extension of the current Web in which information is given well-defined meaning based on some standardized abstract data representation. Currently, this standardized abstract data representation is based on the Resource Description Framework (RDF).

The RDF [53] is a language for representing information about resources on the Web. It is primarily intended for the representation of metadata about Web resources, such as the title, author and modification date of a Web page, copyright and licensing information about a Web document, or the availability schedule for some shared resource. However, by generalizing the concept of a "Web resource", RDF can also be used to represent information about things that can be identified on the Web, even when they are not directly retrieved on the Web. RDF provides a common framework for expressing this information so it can be exchanged between applications while maintaining the integrity of the intended meaning.

2.7.2 Human-Centric Query Processing

2.7.2.1 Textual Queries

Perhaps the most user–friendly way of formulating a search query is by means of natural languages (NL). For example, expressions such as "Who won the Nobel peace price last year?", "Tell me about internal combustion engines", or "What is the weather like in London today?" are most natural for human users. Natural language processing is the field of research dedicated to the study of machine understanding of natural language expressions by analysing word relationships to identify its implicit phrases, significant terms and overall meaning.

While all major search engines provide some degree of NL query handling, the best known are *Ask Jeeves* [54] and Northern Light [55]. In response to a query, Ask Jeeves parses the sentence and matches the elements to about 25 million questions for which it has the answers. It is best for questions such as "How many countries are there in the world?", which tend to have close-formed answers. It also extracts keywords and submits searches to other search engines.

While most NL processing algorithms tend to remove stop words (e.g. "a", "the"), Northern Light's method does not discard such words. It also does not perform full stemming of words (e.g. leave ↔ left). It is also reportedly able to determine the meanings of certain words in different contexts, such as when "not" is used as an operator or as part of a phrase.

However, the NL processing capabilities of search engines are still limited. Full NL processors can reliably interpret acronyms (e.g. ATM, WHO, ANN), understand concepts such as sets and memberships (e.g. Iberia, Spain and Portugal), and distinguish between words having different meanings in different contexts (e.g. cricket). One of the major advantages of the current NL query technology is that it prompts users to enter more words to describe their topic of interest. For example, the query "how are points scored in a cricket match?" gives the search engine program much more useful information than simply "cricket".

Multilingual query processing is another area gaining importance as more and more non-English Web pages are being made available on the Web. With over 150 countries having access to the Internet, the potential of multilingual processing is enormous. To facilitate multilingual query processing, non-English Web pages must first be indexed. Since indexing amounts to the extraction and storage of key terms (or ideas) that characterize a Web page, textual analysis is performed on English-language pages. In the case of other Latin-based languages (such as French, German and Spanish), similar techniques suffice, as words are also delimited by white spaces.

Processing of other languages, such as Arabic, Chinese and Japanese traditionally relies on the use of language-specific dictionaries, techniques and heuristics. With so many languages represented on the Web, as well as the continuous evolution of modern languages and the Internet itself, using language-specific methods is not the best way forward. Ideally, a multilingual indexing and retrieval algorithm should be generic enough to handle multiple languages, be compute-efficient on the fly, and be adaptable to changes. The first requirement dictates that the algorithm should rely on the minimal amount of language-specific information, such as grammatical rules. The second requirement means that most of the compute-intensive work must be done offline. The last requirement ensures that the algorithm is scalable and maintainable.

Recently, a number of experimental systems are being developed to provide multilingual Web retrieval, a representative example of what is known as *Catfind Chinese* [56]. This system treats a document (in this case Big5-coded Chinese) as a collection of symbols from which indexing terms are extracted using statistical analysis. Another approach is to develop multilingual authoring methods for Web content providers. For example, Basilia *et al.* [57] are developing methods for processing and structuring cross-lingual hypertext links. They apply natural language processing techniques in an attempt to "add value" to the information implicitly embodied in the text.

In the near future, standardization of multilingual character representation is likely to simply the display and processing of different languages. For example, the Unicode system [58] has been developed which assigns a unique code to each character regardless of computer platform and programming language. Many companies in the computer industry, such as Apple, International Business Machines Corporation (IBM), Sun and Microsoft, have already shown interest in this emerging standard.

2.7.2.2 Multimedia Data Processing

With the increasing volumes of multimedia Web contents available on the Internet, effective tools for finding these contents become necessary. There are essentially three main types of multimedia contents of interest: audio (e.g. speech, music), still images and videos. This section briefly highlights advanced technologies for these types of data beyond what has been accomplished (as described in Section 2.5).

Audio content retrieval techniques benefit not only the search of audio Web content, but can also be applied to video search by analysing the dialog/sound track that accompany the video data (in much the same way that textual information such as caption can assist the indexing and retrieval of videos). For example, Speechbot [59] is an experimental system that facilitates indexing and retrieval of speech data available on the Web. The system primarily relies on processing the speech data (through transcoding and recognition), but also utilizes textual information (transcriptions) when available. Currently, the system indexes contents such as talk radio and news shows (e.g. The Charlie Rose Show [60] and PBS Online NewsHour [61]) and conference video recordings. In the latter case, the system does not process the video data, but relies on the sound track for indexing.

Analysis of visual data towards providing a means of high-level description for the events that occur in a video sequence will also become increasingly important. This is because current indexing and retrieval techniques for visual data, which rely on processing low-level visual feature and textual information, have proved largely inadequate. Ongoing research in this area that is likely to benefit indexing and retrieval techniques for visual data in the near future include representation, recognition and classification of facial expressions, for example, [62–64] and body gestures [65].

2.7.3 Intelligent Agents

In addition to using the learning capability of neural networks to aid indexing and retrieval, there has recently been a revival of research in artificial intelligence for other Internet applications. For example, WebWatcher [66] and Letizia [67] serve as an interface between the user and Web browser that attempts to learn the user's browsing behaviour with the aim of being able to suggest further Web pages that might be of interest to the user [68]. Other agents, such as Browser Biddy [69], help users retrieve multiple files that take a long time to interactive download, or to perform downloading overnight.

WebWatcher is a "tour guide" agent for the Web. The agent accepts user inputs on interested topics and follows the user's browsing preferences via hyperlinks. Since WebWatcher is a server-based agent, it can log data from multiple users to continually train itself and update its knowledge base. It can then highlight hyperlinks that might be of interest to the user based on knowledge gained from previous tours.

Letizia is a client-based agent that collects personal information about the user's browsing habits, and attempts to anticipate items of interest by doing concurrent, autonomous exploration of links from the user's current position. The agent automates a browsing strategy consisting of a best-first search together with heuristics inferring user interest from browsing behaviour.

There are also agents that take a more active approach towards finding information, products and services for users. By using rule-based approaches to search and gather large amounts of information on the Web, and distilling that information, these agents effectively serve as information filters [70] for users. In addition, agent technology is also applied to commercial activities, such as electronic trading, which will be discussed in Chapter 6.

References, Links and Bibliography

[1] http://www.google.com or http://www-db.stanford.edu/~backrub/google.html, 2004.
[2] http://www.yahoo.com, 2004.
[3] http://www.excite.com, 2004.
[4] http://www.lycos.com, 2004.
[5] http://www.microsoft.com/presspass/press/2000/Sept00/BoschPR.asp, 2004.
[6] http://www.dejanews.com, 2004.
[7] http://fuzine.mt.cs.cmu.edu/mlm or http://lycos.cs.cmu.edu, 2004.
[8] A. Emtage and P. Deutsch, "Archie—An Electronic Directory Service for the Internet", *Proceedings of the Usenix Winter 1992 Technical Conference*, Usenix Association, Berkeley, CA, pp. 93–110, 1992.
[9] M. Gray, "World Wide Web Wanderer", http://www.mit.edu/people/mkgray/net/, 2004.
[10] M. Koster, "Aliweb—Archie-Like Indexing in the Web", *Proceedings of the First International World WideWeb Conference*, Elsevier Science, Amsterdam, Netherlands, pp. 175–182, 1994.
[11] http://www.euro.net/innovation/Web_Word_Base/TWW1-html/WebRob1.html, 2004.
[12] O. McBryan, "GENVL and WWWW: Tools for Taming the Web", *Proceedings of the Second International World Wide Web Conference*, Elsevier Science, 1994.
[13] D. Eichmann, "The RBSE Spider—Balancing Effective Search Against Web Load", *Proceedings of the First International World Wide Web Conference*, Elsevier Science, 1994.
[14] B. Pinkerton, "Finding What People Want: Experiences with the WebCrawler", *Proceedings of the Second International World Wide Web Conference*, Elsevier Science, 1994.
[15] A. Kingoff, "Comparing internet search engines", *IEEE Computer*, Vol. 30, No. 4, pp. 117–118, April 1997.
[16] M. Koster, "Guidelines for Robots Writers", http://info.Webcrawler.com/mak/projects/robots/guidelines.html, 2004.
[17] S. Chakrabarti et al., "Mining the Web's link structure", *IEEE Computer*, Vol. 32, No. 8, pp. 60–67, August 1999.
[18] http://www.almaden.ibm.com/cs/k53/clever.html, 2004.
[19] http://www.research.digital.com/SRC/personal/monika/papers/sigir98.ps.gz, 2004.

[20] M.R. Henzinger, "Hyperlink analysis for the Web", *IEEE Internet Computing*, Vol. 5, No. 1, pp. 45–50, January-February 2001.

[21] C. Shannon, "A Mathematical Theory of Communication", http://cm.bell-labs.com/cm/ms/what/shannonday/shannon1984.pdf, 2004.

[22] S. Chakrabarti, B. Dom, D. Gibson, J. Kleinberg, P. Raghavan and S. Rajagopalan, "Automatic resource compilation by analyzing hyperlink structure and associated text", *Computer Networks and ISDN Systems*, Vol. 30, pp. 65–74, 1998.

[23] K. Bharat and M.R. Henzinger, "Improved Algorithms for Topic Distillation in a Hyperlinked Environment", *Proceedings of ACM SIGIR Conference on Research and Development in Information Retrieval*, Melbourne, AU, pp. 104–111, 1998.

[24] A. Spink, B.J. Jansen, D. Wolfram and T. Saracevic, "From E-sex to e-commerce: Web search changes", *IEEE Internet Computing*, Vol. 35, No. 3, pp. 107–109, March 2002.

[25] G. Salton, "Developments in automatic text retrieval", *Science*, Vol. 253, pp. 974–979, 1991.

[26] S. Brin and L. Page, "The Anatomy of a Large-Scale Hyper-textual Web Search Engine", *Proceedings of the Seventh International World Wide Web Conference*, Elsevier Science, New York, pp. 107–117, 1998.

[27] http://www.datatrak.net7.co.uk, 2004.

[28] http://www.metacrawler.com, 2004.

[29] L. Introna and H. Nissenbaum, "Defining the Web: the politics of search engines", *IEEE Computer*, Vol. 33, No. 1, pp. 54–62, January 2000.

[30] C. Pu and L. Liu, "Update Monitoring: The CQ Project", *Proceedings of the 2nd International Conference on Worldwide Computing and its Applications*, Tsukuba, Japan, pp.396–411, 1998.

[31] M.D. Rosa, T. Catarci, L. Iocchi, D. Nardi and G. Santucci, "Materializing the Web", *In IEEE Proceedings of the 3 rd IFCIS International Conference on Co-operative Information Systems (CoopIS)*, New York, pp. 24–31, 1998.

[32] L. Liu, C. Pu and W. Tang, "Supporting Internet Applications Beyong Browsing: Trigger Processing and Change Notification", *Proceedings of the 5th International Computer Science Conference (ICSC '99) Special Theme on Internet Applications*, Hong Kong, China, pp. 294–304, December 15–17, 1999.

[33] T. Catarci, "Web-Based Information Access", *IEEE Proceedings of the 4th IECIS International Conference on Cooperative Information Systems (CoopIS)*, Edinburgh, Scotland, pp. 10–19, 1999.

[34] http://mindit.netmind.com/mindit.shtml, 2004.

[35] http://www.tracerlock.com, 2004.

[36] http://www.watznew.com, 2004.

[37] http://www.cc.gatech.edu/projects/disl/WebCQ/, 2004.

[38] H.H. Yu and W. Wolf, "A Hierarchical and Multi-Resolution Method for Dictionary-Driven Content-Based Image Retrieval", *Proceedings of International*

Conference on Image Processing, Santa Barbara, CA, pp. 823–826, October, 1997.
[39] Y. Deng and B.S. Manjunath, "NeTra-V: Toward an object-based video representation", *IEEE Transactions on Circuits and Systems for Video Technology*, Vol. 8, No. 5, pp. 616–627, September 1998.
[40] S.Y. Kung and J.N. Hwang, "Neural networks for multimedia processing", *Proceedings of IEEE*, Vol. 86, No. 6, June 1998.
[41] S.H. Lin, S.Y. Kung and L.J. Lin, "Face recognition/detection by probabilistic decision-based neural networks", *IEEE Transactions on Neural Networks*, Vol. 8, pp. 114–132, January 1997.
[42] S. Debnath, S. Sen and B. Blackstock, "LawBot: a multiagent system for legal research", *IEEE Internet Computing*, Vol. 4, No. 6, pp. 32–37, November-December 2000.
[43] N.H. Lovell, F. Magrabi, B.G. Celler, K. Huynh and H. Garsden, "Web-based acquisition, storage, and retrieval of biomedical signals", *IEEE Engineering in Medicine Biology*, Vol. 20, No. 3, pp. 38–44, May/June 2001.
[44] A. Värri, B. Kemp, T. Penzel and A. Schlögl, "Standards for biomedical signal databases", *IEEE Engineering in Medicine Biology*, Vol. 20, No. 3, pp. 33–37, May/June 2001.
[45] S. Lawrence, C.L. Giles and K.D. Bollacker, "Digital libraries and autonomous citation indexing", *IEEE Computer*, Vol. 32, pp. 67–71, June 1999 Also, http://citeseer.nj.nec.com/
[46] http://www.isinet.com, 2004.
[47] Y. He, S.C. Hui and A.C.M. Fong, "Citation-based retrieval for scholarly publications using KSOM neural network", *IEEE Intelligent Systems*, Vol. 18, No. 2, pp. 58–65, 2003.
[48] http://www.research.digital.com/SRC/virtualpaper/pstotext.html, 2004.
[49] T. Kohonen, *Self-Organizing Maps*, Springer, 1995.
[50] H.L. Vu, S.C. Hui and A.C.M. Fong, "Monitoring scientific publications over the WWW", *The Electronic Library*, Vol. 21, No. 2, pp. 110–116, 2003.
[51] A.C.M. Fong, S.C. Hui and H.L. Vu, "Effective techniques for automatic extraction of Web publications", *Online Information Review*, Vol. 26, No. 1, pp. 4–18, 2002.
[52] http://www.scientificamerican.com/print_version.cfm?articleID=00048144-10D2-1C70-84A9809EC588EF21, 2004.
[53] http://www.w3c.org/TR/rdf-primer/, 2004.
[54] http://www.ask.com, 2004.
[55] http://www.northernlight.com, 2004.
[56] http://www.csis.hku.hk/~catfind/, 2004.
[57] R. Basili, M.T. Pazienza and F.M. Zanzotto, "Web-Based Information Access: Multilingual Automatic Authoring", *Proceedings of the IEEE International Conference on Information Technology: Coding and Computing*, Las Vegas, NV, pp. 548–553, 2002.
[58] http://www.unicode.org, 2004.

[59] http://speechbot.research.compaq.com/, 2004.
[60] http://www.bloomberg.com/tv/crose.shtml, 2004.
[61] http://www.pbs.org/newshour/home.html, 2004.
[62] Y. Guo and H. Zhang, "Facial Expression Recognition Using Continuous Dynamic Programming", *Proceedings of the IEEE ICCV Workshop on Recognition, Analysis, and Tracking of Faces and Gestures in Real-Time Systems*, Vancouver, Canada, pp. 163–167, 2001.
[63] S. Akamatsu, J. Gyoba, M. Kamachi and M. Lyons, "Coding Facial Expressions with Gabor Wavelets", *Proceedings of the Third IEEE International Conference on Automatic Face and Gesture Recognition*, Nara, Japan, pp. 200–205, April 14–16, 1998.
[64] M. Pantic and L.J.M. Rothkrantz, "An Expert System for Multiple Emotional Classification of Facial Expressions", *Proceedings of the 11th IEEE International Conference on Tools with Artificial Intelligence*, Chicago, Illinois, pp. 113–120, 1999.
[65] L. Bretzner, I. Laptev and T. Lindeberg, "Hand Gesture Recognition Using Multi-Scale Colour Features, Hierarchical Models and Particle Filtering", *Proceedings of the Fifth IEEE International Conference on Automatic Face and Gesture Recognition*, Washington, DC, pp. 405–410, May 20–21, 2002.
[66] http://www-2.cs.cmu.edu/afs/cs.cmu.edu/project/theo-6/webagent/www/project-home.html, 2004.
[67] http://lieber.www.media.mit.edu/people/lieber/Lieberary/Letizia/Letizia.html, 2004.
[68] D.E. O'Leary, "The internet, intranets, and the AI renaissance", *IEEE Computer*, Vol. 30, No. 1, pp. 71–78, 1997.
[69] http://www.doeasier.org/browserbuddy/, 2004.
[70] http://cmc.dsv.su.se/select/information-filtering.html, 2004.
[71] R. Filman and Feniosky Peña-Mora, "Seek, and ye shall find", *IEEE Internet Computing*, Vol. 2, No. 4, pp. 78–83, July-August 1998.

CHAPTER 3

THE INTERNET AS A COMMUNICATIONS MEDIUM

3.1 INTRODUCTION

The Internet is fast becoming a popular communications medium due to its continuous availability and wide geographic coverage. It also provides a means to circumvent the relatively high charges of long-distance telephone calls via the traditional circuit-switched network. Already, many modes of human communications, such as the written word (text), speech and videoconferencing, are supported.

The ubiquitous electronic mail (e-mail) is perhaps the most widely used method of person-to-person communication via the Internet. Originally used in the realms of the selected few (e.g. academic institutions and military installations), e-mail is now used by the masses. It is ideal for the delivery of textual information. Nowadays, it is also possible to send multimedia data files, such as video clips and images, as attachments together with e-mail messages. This makes e-mail more popular than ever.

In a wider sense, e-mail can be considered a special instance of what can be described as messaging services. Other examples of messaging services include fax, voicemail and short messaging services (SMS), all of which can be effectively delivered via the Internet. These messaging services are characterized by the lack of real-time information delivery. In some situations, these messaging services are preferable to real-time services. For example, a recipient might be too busy to answer a telephone call, preferring to decide whether to respond to a voicemail at his own convenience.

Real-time communication services, on the other hand, provide modes of communication that are more natural to human beings. These services are characterized by the fact that transmission of information is performed instantaneously, or near-instantaneously with a delay (between the time the message is sent and the time it is received) that is typically imperceptible. Obvious examples include Internet telephony, instant messaging, online voice or text chat and videoconferencing.

While using the Internet as a communications medium offers many advantages compared to traditional means, there are many challenges involved in ensuring a high quality of service (QoS) for both real-time and non–real-time services. Obviously, the performance of any Internet-based communication system must be good enough to rival (or better) traditional approaches to achieve success in the marketplace. The solution must also be cost-effective allowing for savings to be passed to the consumer.

The Internet differs from the traditional circuit-switched network such as public switched telephone network (PSTN) in that there is no fixed physical channel between the sender and receiver during the course of transmission. Instead, a message (which may be (portions of) text, image, video frame, speech signal, and so on) must first be digitized and packaged into packets to be sent via *ad hoc* channels, which is likely to change during the course of transmission of the packets that make up a message. The receiver must then reconstruct the message using the packets received. A fundamental problem with the current best-effort Internet (Internet Protocol (IP) version 4 or IPv4) is that it does not provide any QoS guarantees. Due to limited bandwidth resources and other environmental factors such as noise, packets may be lost or received in a different order from what was transmitted.

For real-time applications, such as video streaming, delay (especially varying amounts of delay) can cause as much trouble as outright packet losses because data that arrive after their original intended playout time usually have to be discarded. Although the emerging IP version 6 (IPv6) includes provision for QoS control (in addition to solving the addressing problem), there is much uncertainty regarding its widespread acceptance by the public. The Internet will therefore remain predominantly IPv4 with islands of IPv6 networks here and there [1, 2].

This situation calls for innovative solutions that will not only tackle the best-effort nature of the current Internet, but other related problems such as network heterogeneity, receiver heterogeneity and different needs for different applications. Since the Internet is nothing more than a very large connection of networks, network heterogeneity must be expected. It refers to the fact that while parts of the Internet are made up of state-of-the-art networks, others are mediocre or even very old and poor. Data that are sent from different parts of the Internet will then be subjected to varying degrees of treatment.

Receiver heterogeneity refers to a similar problem as network heterogeneity, but applies to receivers. Obviously, there is much difference in the processing power between proper workstations or mainframe computers compared to personal computers (PCs) or even legacy systems that lack both processing power and local storage. Nowadays, with the proliferation of handheld communications devices (e.g. cellular phones, wireless capable personal digital assistants (PDAs), etc.), the disparity between top-of-the-range receivers and bottom-end receivers will only widen.

Depending on the mode of access to the Internet, the effectiveness of Internet-mediated communications can vary. For example, if 56k dialup modem is used to access the Internet, then the quality of high-bandwidth applications (such as voice and video) is likely to be poor compared with, say, cable Internet access.

Different services obviously have different requirements on the use of channel resources. For example, real-time applications such as Internet telephony and video streaming require a strict timing regime with minimal delay, which may be achieved at the expense of some loss of fidelity. On the other hand, transmission of non–real-time text data, for example, is more tolerant to delay but often requires a high level of data integrity in the received messages.

This chapter presents various aspects of using the Internet as a communications medium that is used to support various modes of human communications. In particular, the following topics are covered:

- The various protocols used for Internet-mediated communications, such as IP, transmission control protocol (TCP), real-time transport protocol (RTP), user datagram protocol (UDP).
- Electronic mail.
- Online presence notification and instant messaging. Online presence notification gives information on whether a user is online or not and is therefore important for users to facilitate real-time communications, such as instant messaging.
- Internet telephony.
- Video data transmission.
- Videoconferencing.
- Unified messaging.

3.2 Internet Communication Protocols

The role of Internet communication protocols [3, 4] in the transmission of multimedia data is summarized in Figure 3.1. Two transport protocols, TCP [5] and UDP [6], sit on top of IP [7]. TCP provides a reliable stream-oriented delivery service with end-to-end error detection and correction, while UDP provides connectionless datagram delivery.

RTP [8], which is designed for real-time data transmission, usually runs over UDP/IP but can also run over other protocols such as TCP/IP. Two popular application-level protocols for the World Wide Web (WWW) are the Hypertext Transfer Protocol (HTTP) [9, 10] and Real-Time Streaming Protocol (RTSP) [11].

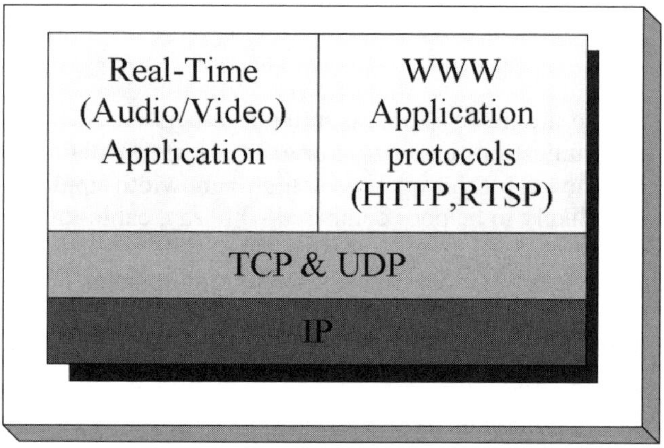

Figure 3.1 Internet communication protocols

3.2.1 Transmission Control Protocol (TCP)

TCP is a connection-oriented protocol and provides for the reliable transfer of data over the Internet. TCP divides information into smaller data packets before transmission. These data packets are then reassembled into the original information at the receiving side. The receiver needs to generate a checksum for each packet and compares it to the checksum included in the header of data packet to verify its correctness.

In the event of a failure (error found in packet received), the receiver requests for retransmission of the data packet by the sender until a successful transmission is attained. TCP can provide a guarantee for data communication over the Internet by using a handshaking mechanism to ensure reliable data transfer. However, extra overhead is incurred in the generation and maintenance of a reliable connection. In particular, the latency incurred by the retransmission process is often excessive for real-time applications, such as video streaming. TCP is, therefore, most suitable for applications that can tolerate a delay in transmission but require the utmost data integrity in the delivered messages.

3.2.2 User Datagram Protocol (UDP)

UDP provides unreliable data delivery over the network with minimal overheads. UDP does not provide mechanisms to ensure guaranteed data delivery. Hence, data may be lost, may arrive out of order, or duplicated. However, the low overhead in UDP transmission makes it very efficient for real-time data transmission over the Internet.

3.2.3 Real-time Transport Protocol (RTP)

RTP provides end-to-end delivery services for data with real-time characteristics, such as live audio, video and simulation data. These services include payload type identification, sequence numbering, time-stamping and delivery monitoring. Since RTP does not address resource reservation and does not guarantee QoS for real-time services, it typically runs on top of UDP to make use of its multiplexing and checksum services, and both protocols contribute parts of the transport protocol functionality. RTP consists of two closely linked parts: RTP Data Transfer Protocol and RTP Control Protocol (RTCP).

3.2.3.1 RTP Data Transfer Protocol

RTP Data Transfer Protocol carries data that have real-time properties. It defines basic packet format to support real-time communication but does not define control mechanisms or algorithms. This protocol is often integrated into the application's processing operations rather than implemented as a separate layer. The packet formats provide information required for audio and video data transfer, such as incoming packet sequence number of video data. Sequence numbers can be used for packet loss detection or determining the position of a frame within a video sequence.

3.2.3.2 RTP Control Protocol (RTCP)

RTCP monitors the QoS and control information related to an RTP data stream. The most important predefined control information elements are sender and receiver reports, which are generated by the source and the destination, respectively. Both contain information about the current QoS of the data stream and receiver blocks. Each receiver block contains receiver statistics about the data stream such as the highest sequence number received, fraction of lost packets since the last report, and cumulative number of lost packets. The sender report also includes sender information such as Network Time Protocol (NTP) and RTP time stamps and the number of packets and bytes already transmitted. A receiver uses this information to calculate receiver block statistics.

3.2.4 Hypertext Transfer Protocol (HTTP)

HTTP is an application-level protocol for distributed, collaborative, hypermedia information systems. It runs over TCP or any reliable transport protocol. HTTP employs a client-server mechanism. A Web browser (client) sends its request to a Web server and the server then responds to it. For example, a user can use a browser to ask a Web server for information using the GET request method, or PUT transfers information from client to server.

Each request requires its own TCP connection. The TCP connection closes after the client and server finish exchanging the request and response messages. This approach is obviously inefficient, since each new TCP connection costs one round trip in time [4]. HTTP 1.1 now permits a single TCP connection for several HTTP transfers by defining additional methods such as OPTIONS, PUT, DELETE and TRACE.

3.2.5 Real-Time Streaming Protocol (RTSP)

RTSP is a client-server multimedia presentation control protocol. RTSP is designed for the efficient delivery of streamed multimedia over IP networks.

RTSP provides an extensible framework to make controlled, on-demand delivery of real-time data possible, such as audio and video. Sources of data can include both live audio/video and stored audio/video clips. This protocol is developed to control multiple data delivery sessions, provide a means for choosing delivery channels such as TCP, UDP and multicast UDP, and provide a means for choosing delivery mechanisms based upon RTP. The protocol supports VCR-like functionality, such as play, fast-forward, pause and stop. Also, it understands time stamps, which are used for time-based search within a presentation. It also supports multicasting for simultaneously broadcasting multiple data files.

3.2.6 Illustration

Figure 3.2 illustrates an example of using the above Internet communication protocols for real-time video transmission. First, only control information about the real-time data to be transferred are exchanged between the Web client and Web server using HTTP/TCP. The Web browser then launches a helper program such as a browser plug-in with the parameters retrieved from the Web server. These parameters may include the video data's encoding type [such as Moving picture experts group (MPEG) video or Motion Joint photographic experts group (JPEG) video], or the video server's address to contact for retrieving the video data.

Next, the video helper program and server run an RTSP protocol to exchange control information required to start and stop data transmission. It runs over TCP or UDP and is not connection-oriented. The video data are received by a video client program, which may be identical to the helper program or run on a separate machine. The client program can then play out the video as it arrives at the client. The data are transferred using a protocol other than RTSP or HTTP, for example, RTP/UDP. RTSP can negotiate the transfer protocol with the video server.

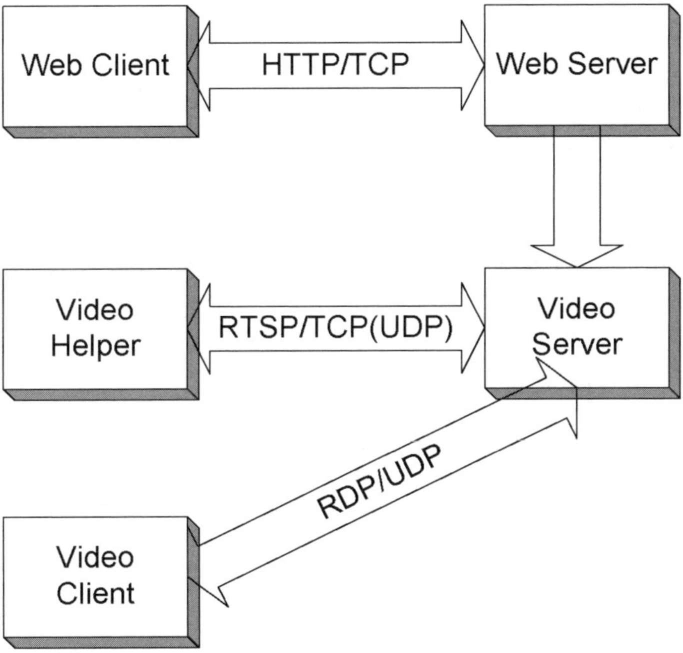

Figure 3.2 An application of Internet communication protocols to the transmission of real-time video data

3.2.6.1 Discussion

TCP is not really suitable for the transfer of real-time data such as audio and video except for some special situations (e.g. excellent network conditions) such that the round-trip-time for retransmission is bounded to within the allowable tolerance. To provide reliable and sequenced service, TCP recovers packet losses by retransmissions. When delivering real-time data, waiting for a receipt of all retransmissions introduces intolerable delay in case of packet losses. UDP avoids this drawback with simple datagram-oriented (but unreliable) service.

RTP and its associated RTCP, however, support real-time data transfer. Therefore, RTP, along with UDP, can be used for video transmission. However, TCP may be more appropriate for exchanging non–real-time data such as control information since it guarantees the delivery of every data packet.

Finally, RTSP is an application layer protocol for the control of streaming data on the Internet. RTP and RTSP can be used together to enable the efficient delivery of real-time video over the WWW.

3.3 ELECTRONIC MAIL

The electronic mail (e-mail) has become an increasingly important way of communication with the increasing popularity of Internet. In fact, e-mail has grown from a mere convenience to a mission-critical messaging application [12]. It is the backbone for collecting and disseminating corporate information and for crucial communications both within and outside a company. Today, e-mail is complementary to more traditional forms of communication that are widely used, such as conventional mail, telephone and fax. Using e-mail, individuals are able to correspond quickly and efficiently even if they are physically separated by long distances.

3.3.1 E-mail Protocols

The Simple Mail Transport Protocol (SMTP) [13] has been the most popular mail transport protocol used. Initially, e-mail messages were mainly text-based. In recent years, however, there is an increase in sending non–text-based attachments such as pictures, audio/video clips and binary files to e-mail messages. The Multipurpose Internet Mail Extensions (MIME) [14, 15] format was implemented to address the packaging of mail attachments.

Since SMTP is a 7-bits text delivery protocol and cannot handle non-textual characters, binary attachments need to be converted to text characters for transmission over SMTP. The base-64 encoding scheme [16] is used for the conversion. As a result, e-mail messages with attachments often increase in size up to about 40%. This increase in mail size is not a very important issue for wired networks but will degrade the performance of wireless networks, which are typically characterized by having very limited bandwidth resources.

Besides the increase in mail size due to attachment conversion, another problem is that the size of the mail attachments themselves are getting bigger. This is due to the trend of e-mail users sending bigger image, audio, executable or even video file attachments with their e-mail messages.

In the past, e-mail systems delivered mail messages to a destination workstation. Unfortunately, this mode of operation presents problems if a user frequently changes workstations or perhaps shares a workstation with other users. Today, most e-mail systems deposit mail in mailboxes located on some types of e-mail server systems. The Post Office Protocol (POP) [17] is a protocol designed to let users login to a POP server and retrieve mail from a remote system rather than login to the network itself. POP enables the user to access his mailbox from any system on the Internet. Therefore, the Internet mail transport system consists of two components: SMTP and POP. In other words, SMTP is used for sending mail while POP is used for the retrieval of mail.

3.3.2 E-mail Systems

There are many electronic mail systems available. Among these systems, SMTP is still the most popular mail traffic protocol used. However, SMTP supports only rudimentary text format. In addition, SMTP only delivers the mail messages. It does not guarantee delivery, issue a return receipt, allow "unsending" or carry attachments. Despite all of these shortcomings, SMTP has worked reliably for over 20 years and has been supported by virtually all major messaging systems.

3.3.2.1 Proprietary E-mail Systems

Apart from the Internet SMTP/POP mail system, there are other proprietary mail systems such as Microsoft Exchange [18] and Lotus Notes [19]. These proprietary mail systems provide additional features such as delivery notification, on-demand retrieval, mail filtering, mail routing, calendaring and scheduling.

These mail systems are highly integrated and require their own proprietary mail clients to access the mail services. In order to communicate with other Internet mail users, these systems must integrate the basic Internet mail protocol natively into their systems or implement gateways for mail exchange. These proprietary mail systems can operate in a wireless network environment, but do not generally include mechanisms that cater for the special characteristics of wireless networks.

3.3.2.2 Web-based E-mail Systems

Currently, there are many free web-based e-mail systems available. The more popular ones include Hotmail [20] and Excite mail [21]. These e-mail systems also provide enhanced features such as delivery notification, on-demand retrieval, mail filtering, mail routing, calendaring and scheduling. However, as in the case of proprietary e-mail systems, certain features such as delivery notification are only valid for mail sent within the system. E-mail messages sent out to users using other e-mail systems will not receive delivery confirmation or acknowledgement. These e-mail systems are accessible via web browsers such as Internet Explorer and Netscape. Therefore, they can also be supported in a wireless network as long as Web service is available on the wireless network.

Unlike proprietary e-mail systems, these Web-based e-mail systems usually serve the public Internet users instead of users within a corporate environment. These Web-based e-mail systems generate revenues for their providers via advertisements and sponsorships. Some proprietary e-mail systems such as Microsoft's Exchange also offer Web-based access as front–end-user interface as an alternative to their e-mail clients.

As a result of the large user base, Web-based e-mail systems are often slow and might even be unavailable at times. They also tend to impose relatively strict quotas on the amount of memory for each individual user. This often means that relatively large files (like a few MB) may not reach the mailboxes of Web-based e-mail users. In addition, Web-based e-mail systems have extra overheads. The Hyper Text Markup Language (HTML) [22] code and web advertisements that are delivered together with each mail retrieval operation add to the amount of data transfer. Also, Web-based e-mail systems do not support the local caching of retrieved mail messages. As a result, mail messages have to be retrieved again every time a user wants to view them.

3.4 Online Presence Notification and Instant Messaging

3.4.1 Current Online Presence Notification Approaches

In order to have point-to-point real-time communication with other online users, each user needs to be aware of other online users. This section discusses the techniques for online presence notification.

3.4.1.1 Exchange Server Approach

In this approach, the exchange server (ES) as shown in Figure 3.3 is a dedicated server that maintains and manages the information of active registered users. It acts as an exchange for finding and connecting users. When a user is online, an agent program running on his computer will sign on to the ES. When the user goes offline, the agent program will also sign out the user from the ES. In order to ensure that the list of online users is current and up-to-date, the ES will periodically check each signed-on user to ensure that the user is still online. This is necessary to cater for any abnormal disconnection from the network of signed-on users, and is a particularly important issue for users who use dialup modems to access the Internet. Signed-on users that are unreachable by the ES will be removed from the online user list.

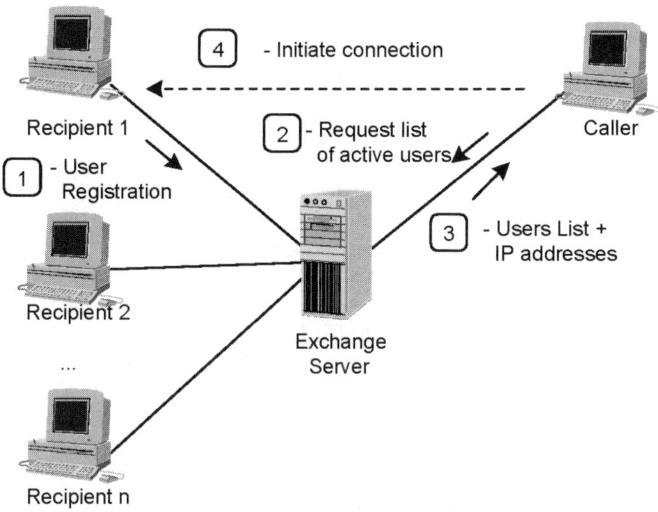

Figure 3.3 Using an exchange server to maintain users' online status

3.4.1.2 Electronic Mail Approach

This method uses fixed e-mail addresses as a means to check the online presence of other users. As shown in Figure 3.4, the requester will only need to know the e-mail address of the recipient. The SMTP protocol is used to transfer a special e-mail consisting of the requester's IP address and other information to a SMTP mail server, which will send the mail to the recipient's mailbox.

At the recipient's end, an agent program will poll his mailbox using the POP protocol periodically for such special mail. Upon the reception of such special mail, the requester's IP can be extracted and a communication session can be initiated by the agent according to the type of communication mode specified in the special mail. A time-out period is also defined in the special mail to indicate the period of validity of the call request.

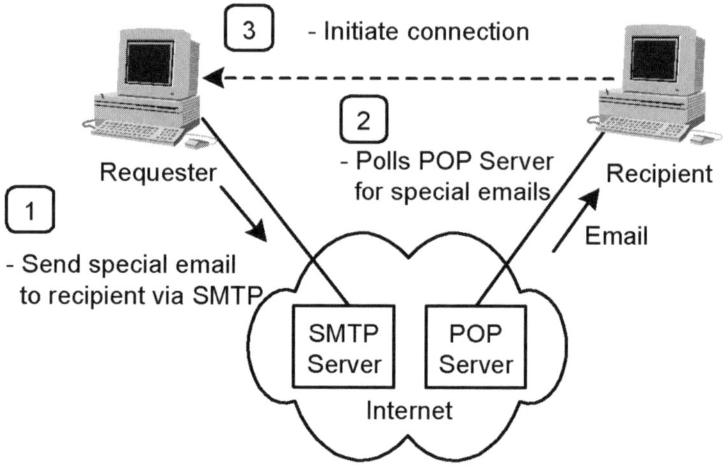

Figure 3.4 Using e-mail to maintain users' online status

3.4.1.3 Discussion

Both the ES and the e-mail approaches allow the notification of online presence to users. However, the ES approach has more advantages as the inclusion of a dedicated ES offers flexibility and control in system implementation. Also in the e-mail approach, special notification e-mails might be accidentally removed by the user's mail client programs.

There are two variations of implementation using the ES approach. In the first case, the ES mainly authenticates and manages users and provides online presence information. Person-to-person communication such as instant messaging is done directly between users. As such, the system is very prone to various attacks such as message spoofing and anonymous messaging. In the second case, all messaging is done via the server. This approach provides more security but it increases the load for the server.

In the adoption of the ES approach, there are different ways to provide online presence information. It can be a public list type where all current signed-on users are visible to one another, or a subscription list type where only authorized associates will be notified of the online presence of the user. The public list organization allows more interaction to take place as signed-on users can view each other. Depending on the structure of the ES, the organization of the user list can even be partitioned into location, language, interest, and so on to facilitate communication between users with commonality. However, such exposure might result in voluminous interaction requests from strangers. The subscription list organization prevents unsolicited interaction requests from strangers since only authorized signed-on users will be notified of the online presence of the user.

3.4.2 Instant Messaging Systems

Instant Messaging Systems (IMS) allow online users to send messages to one another in real time. Client software installed on computers informs users when any individuals in their list of "buddies" (e.g. colleagues, workgroup members, friends, and so on) logon to the network so that they can communicate with each other. It also notifies them if a "buddy" sends a message to them.

3.4.2.1 Some Popular Public Instant Messaging Systems (IMS)

Table 3.1 summarizes of the characteristics of some popular public IMS.

Table 3.1 Characteristics of popular public IMS

System	Model	Protocol	Listing type	Listing storage
ICQ	ES	Proprietary	Subscription	Client side
AIM	ES	Proprietary	Subscription	Client side
YAHOO!	ES	Proprietary	Subscription	Server side
MSN	ES	Proprietary	Subscription	Server side

ICQ, I Seek You; AIM, AOL Instant Messenger; **ES**, exchange server.

ICQ [23] stands for "I Seek You". It is an instant messenger program invented by Mirabilis (a software company, which was later sold to America Online (AOL) for about US $400 million in 1998). Registered users are allocated a Unique Identification Number (UIN) and allowed to see whether their friends in their contact list are online, and to communicate with them. Users can send text messages, Universal Resource Locators (URLs), greeting cards, voice messages, chat requests and transfer files ICQ has over 38 million signed-up users, which makes it one of the world's largest Internet online communication networks. It is officially available for all versions of Windows and Mac operating systems. For the Linux environment, a cloned client program, LICQ [24] is also available although the legitimacy of the clone program remains unclear.

While ICQ is an innovative and useful program, it is also quite insecure. This is because too many operations are performed by the client program (as client-side operations). Client-side operations make ICQ vulnerable to attacks. For example, it is possible to spoof messages (i.e. send fake messages that will appear to be sent from a different user) on ICQ. This is because ICQ will receive messages from every IP address. Since ICQ simply accepts messages from anyone instead of from the server only, it is possible to engineer a fake message and send it to the victim. Another example of its weakness is authorization tampering. ICQ allows users to require authorization before other users can add them into their contact list. However, since the client program does the verification, it is possible to use patching code that is readily available from the Web to patch the client program such that anyone can be added into the contact list without permission.

The AOL Instant Messenger (AIM) [25] is also a subscription list application that provides presence notification and instant messaging capabilities. Provided free of charge by America Online Inc, it has over 30 million registered users. Users register themselves using Screen Names and add their friends to their Buddy List. Unlike many other IMS, AIM users can add one another to their buddy lists without approval. In this case, users always know when their buddies are online. Unlike ICQ, MSN Messenger and Yahoo! Messenger, AIM does not provide an integrated free e-mail account, and thus, users cannot send or check mail directly from the client. However, just as with Yahoo! Messenger and ICQ, users can send files to anyone on their buddy list. AIM also lets users share folders with one another.

Yahoo! Messenger [26] is an ambitious IM client that offers e-mail integration, file transfer, group chat, voice conferencing and message alerts. Users need to have a free Yahoo! account to download and install the messenger. However, unlike AIM and MSN Messenger, users are not restricted to using Yahoo!'s free e-mail account. User profiles are stored on Yahoo!'s central database. Yahoo! Messenger allows file transfer, but there is a 500 kb limit to the file size. On the other hand, it does not support client-to-client direct file transfer, so all files are uploaded to Yahoo! prior to final delivery. Unlike AIM, users cannot set up a shared folder. Like other IM clients, Yahoo! Messenger is not interoperable with other IM clients and its protocol is proprietary and classified. But Yahoo! Messenger's tight links with personalized Yahoo! Web products enable it to attract numerous Yahoo! users.

MSN Messenger [27] integrates with Microsoft's Web-based e-mail service, Hotmail. Unlike Yahoo! Messenger and ICQ, which let users choose their e-mail client, MSN Messenger only lets users send e-mails via Hotmail. Like Yahoo! Messenger, subscription lists are stored on MSN's servers, so that users can access them from any machine. While some functions, such as file transfer, are not supported, MSN Messenger provides support to invite other users to an online game or NetMeeting session.

MSN Messenger also attempts to interoperate with AIM. However, AOL has blocked users that try to login to its network using MSN Messenger as AOL considers it a security risk to its users. This lack of interoperability means that users have to run many IM client software on their computing device if they have friends from different IM networks. IM solution providers that have large user bases are reluctant to participate and support open protocol design as it would provide interoperability and eventually dilute their user base.

3.4.2.2 Discussion

The above survey highlights many interesting and useful features provided by the popular IMS. However, they also have common deficiency due to their attempts to fight for a user base. They all keep their protocol confidential and they do not want to provide interoperability with other vendors.

The idea is that there are some third party vendors (not involved in the development or production of the original IMS) produce clone/generic/interoperable clients that are compatible with the OEM IMS protocols. These 3rd party vendors achieve this by monitoring and analysing network packets that leave the OEM systems from the IM client they (these 3rd party vendors) have tested.

Besides competition issues, another valid reason for keeping their protocols secret is security. While most IMS are generally secure and offer lots of security features, their system integrity might be compromised if their protocol were made public. This is because malicious programmers can create a non-conforming IM client that exploits the server and undermines user security if they know the protocol. Therefore, to design an open protocol to provide interoperability where the security and integrity of the system will not be compromised is a challenging task.

Most IMS are not restricted only to instant messaging services, but also provide a means for the initiation of real-time chat services and files transfer. These additional services can only be incorporated into the client by the vendors themselves. Therefore, it would be helpful to implement a generic online notification system that has a communication service negotiation framework such that third party developers can add new communication services into the IMS easily and legally.

3.4.3 The Online Presence Notification Protocol

In a bid to promote interoperability, the authors have developed an open and secure protocol known as the *Online Presence Notification Protocol* (OPNP) and have implemented an experimental system (Online Presence Notification System (OPNS)) based on the OPNP [28]. OPNP is designed to support online presence notification as well as negotiation for real-time communication services to online users. It moves beyond instant messaging by supporting service negotiation so that OPNP-compliant systems can also provide other types of services such as text-based chat, voice chat, videoconferencing and direct file transfer. The protocol is designed to avoid the flaws and security weaknesses of current IMS. It prescribes an open standard that facilitates interoperability and provides the system designer with some scope for customization.

3.4.3.1 Architecture Model

The layered architecture model of the protocol is shown in Figure 3.5. The topmost layer in the model consists of the server, client and real-time communication applications.

The server acts as an ES to provide online presence information to users via subscription list model. It also acts as an exchange for real-time communication service negotiations between users. A user's subscription list and settings are stored at server side so that the subscription list and settings appear the same regardless of which computer the user uses. The server replicates such information as a means to build redundancy into the system to enhance robustness. The Internet e-mail address is used as the user ID, as its naming space is infinite, globally unique and human-readable. To protect the user's privacy, the subscription of users to the subscription list must be authorized by the subscribed users. Online presence information will only be revealed to authorized subscription. Users may block other users from incoming communication requests at will. In addition, users are allowed to revoke authorization given so that they need not maintain a large blocking list.

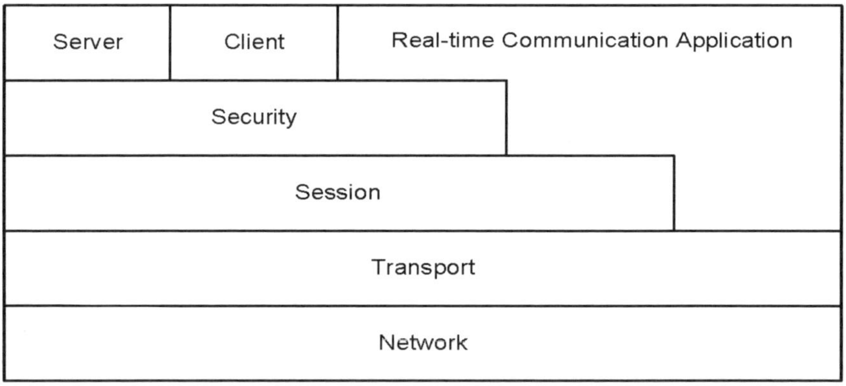

Figure 3.5 Layered architecture model for the OPNP

The client program provides a means for the user to connect to the server and to launch real-time communication modules for services such as instant messaging, text-based chat, voice chat, and so on. The client program does not provide any real-time communication services directly, but it provides negotiation support via the server for users to initiate real-time communication services with other online users in the subscription list. Along with proper encryption and session identification schemes, this arrangement eliminates fake requests, as the client application will only listen for communication requests via the server. Service requests will not be made in situations where the recipient user is not online. Upon successful negotiation, real-time communication will be performed directly between communication modules. This ensures that the server is not overloaded and ensures good scalability.

The Security Layer provides data encryption support using session keys. A hybrid cryptographic authentication protocol is used so that the password is encrypted before it is delivered over the network. This prevents password sniffing to avoid impersonation of users.

The Session Layer supports session identification. Communication connections between client and server use an ordered incremental session ID scheme for session identification. An increment constant is used to increase the value of the session ID for each connection established between the client and the server. Communications between the client and the server will be recognized based on the proper increment of the session ID.

For the Transport Layer, communications between clients and servers use short-lived TCP/IP. TCP/IP provides reliability for data delivery. Since data transfer for online presence notification and service negotiation are sporadic and short in nature, short-lived TCP connections are suitable for this purpose. Communications between real-time communication modules can use TCP/IP, UDP or RTP. Communication module developers are free to implement or use any transport mechanism that is IP compatible.

Finally, the Network Layer in the model includes any computer network that supports the IP protocol. It includes Intranets formed by local area networks (LANs) and the Internet.

3.4.3.2 Security Features

The hybrid cryptographic scheme in this research uses public key encryption [29] to distribute session keys and uses the distributed session keys to encrypt data using symmetric encryption. Therefore, the hybrid encryption scheme offers an improvement in performance without compromising security. Session keys are no longer transmitted over the network once they are securely distributed by public key encryption schemes. In addition, the session key is only valid for the service session and is unique between each user pair. The session keys are then used for encrypting service requests and messages.

3.4.3.3 Communication Processes

OPNP uses a simple but flexible packet structure. Besides the header marker, only the first two fields, version number and user ID, are fixed in order. The subsequent packet content comes in the form of "=". Each field is separated by a one-byte delimiter. The protocol supports versioning to provide extensibility. By allowing the exchange of protocol version information, new features can be incorporated without out-dating the protocol. Figure 3.6 shows the communication processes between the client and server via short-lived non-persistent TCP connections over the network. The functions of the processes are described below.

- *Sign-on process* — this is used to perform authentication on users who wish to use the service.
- *Session-keep-alive process* — after the user has signed on, the client sends keep-alive packets to the server periodically. In the absence of keep-alive packets, the server will sign off the user and invoke the *Notification Process* to alert other subscribers.
- *Notification process* — when a user signs on/off or changes his online status, the server informs other signed-on users subscribed to the presence information of the user.
- *Service negotiation process* — this process enables signed-on users to request for real-time communication services with each other.
- *Subscription process* — this allows users to maintain their notification subscription list.
- *Administration process* — this allows users to update their particulars, change preferences and to block, delete or revoke subscription authorization of users subscribed to the presence information of the user.
- *Sign-off process* — besides a normal sign off from the service, this process is useful when the user has been abnormally disconnected from the network and dynamically assigned a new IP address.

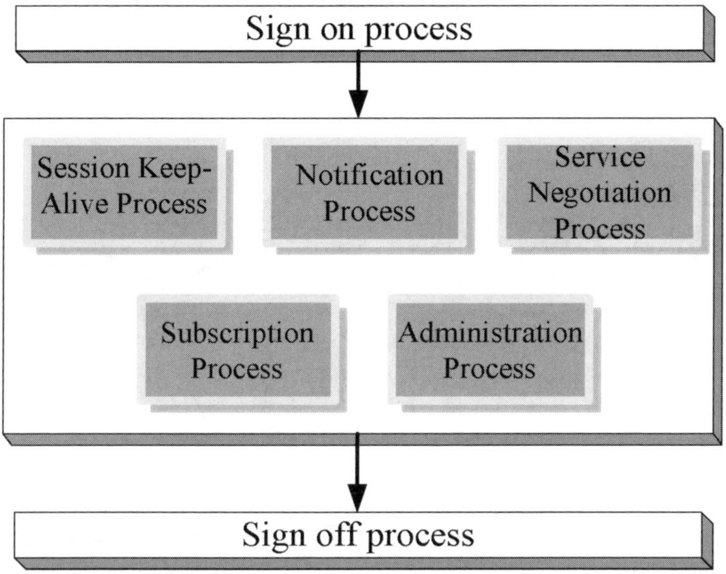

Figure 3.6 OPNP communication processes

3.4.4 Online Presence Notification System

Based on the protocol described in Section 3.4.3, the authors have developed an OPNS. It adopts the ES approach using TCP/IP as the transport protocol to provide presence notification service. OPNS also provides a platform for initiating various kinds of real-time communication services.

3.4.4.1 System Architecture

Figure 3.7 shows the system architecture of OPNS. The OPNS server consists of seven main components:

- *Administration Engine* — this processes user registration and provides user profile and subscription list management.
- *Authentication Engine* — this handles users' sign on. It also generates and issues session keys, session IDs and increment constants to signed-on users.
- *Session Engine* — this monitors the online status of each signed-on user. It updates and manages the online user database to reflect the current online status. Time-out mechanism is included to handle the situation of abnormal termination of signed-on users.

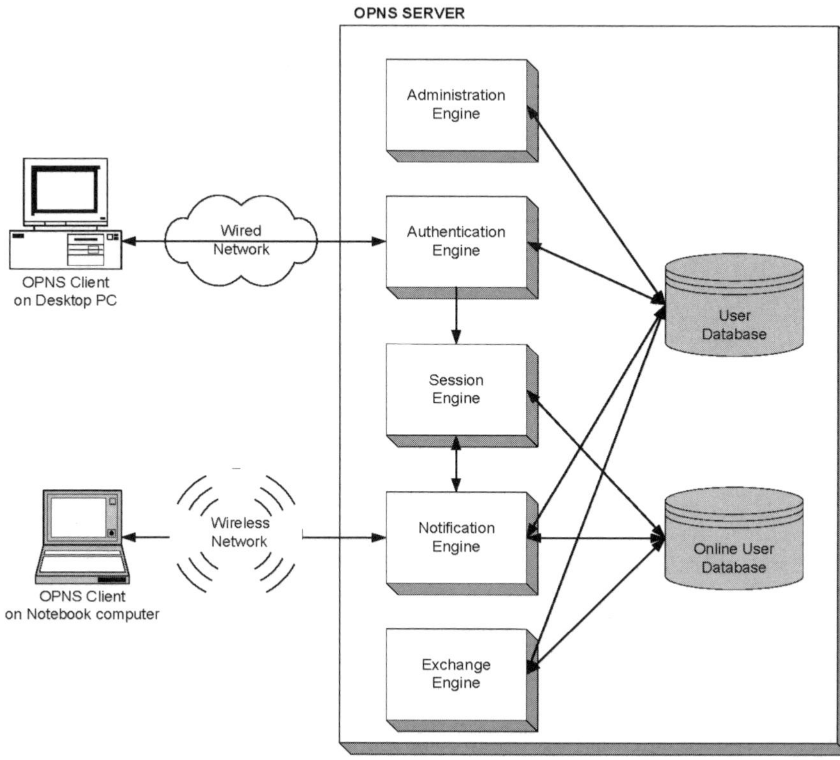

Figure 3.7 OPNS architecture

- *Notification Engine* — this notifies signed-on users the presence of other online signed-on users that are in their subscription list.
- *Exchange Engine* — this provides a means for negotiation for real-time communication services between concurrently signed-on users.
- *User Database* — this database, which is managed by the *Administration Engine*, stores the ID, password, particulars, preferences and subscription list of all registered users.
- *Online User Database* — this database, which is managed by the *Session Engine*, stores the current list of signed-on users, the session key, session ID, increment constant, IP address and online status of signed-on users.

The OPNS client connects to the OPNS server to sign on the user, and displays the list of online users in the user's subscription list. The user can then select various real-time communication services with other concurrently signed-on users. The OPNS client then initiates a service request to the recipient via the OPNS server. Once informed by the OPNS server that the recipient has accepted the service request, the OPNS client will activate the selected real-time communication module to establish direct communication with the recipient. The OPNS client also provides administrative functions such as the subscription of users to online presence notification subscription list and the updating of a user's particulars and preferences.

3.4.4.2 Comparison of Online Presence Notification System (OPNS) with other Systems

Table 3.2 summarizes the features of the OPNS in comparison with the four popular IMS described in Section 3.4.2. Since OPNS is a generic presence notification system, comparisons are mainly made for system features instead of service features.

Although OPNS shares many system features with other IMS, it has particularly strong security and privacy features. This is made possible by the separation of online presence from communication services and by achieving a fine balance between server and client operations. In the context of OPNS (and therefore OPNP), having strong security features is not just desirable but necessary. This is because an open protocol would likely be subjected to severe security attacks.

While all the flaws in other IMS might be resolved by the correction of protocol and system design, security concerns and the monopoly of a user base have led to the safeguarding of these IM protocols by IMS providers. In contrast, OPNP-compliant systems are relatively safe, even though the protocol is made known publicly to promote interoperability.

Table 3.2 Comparison of OPNS with other systems

Feature	**OPNS**	**ICQ**	**AOL**	**MSN**	**Yahoo!**
System design					
Transport protocol	TCP	TCP/UDP	TCP	TCP	TCP
Notification control	Yes	Yes	Yes	Yes	Yes
Online status customization	Yes	No	No	No	Yes
User ID/name space	E-mail	Unique number	Screen Name	Hotmail	Yahoo! ID
Web-based access	Yes	Yes	Yes	No	Yes

Feature	OPNS	ICQ	AOL	MSN	Yahoo!
User privacy					
Subscription list storage	Server	Client	Client	Server	Server
Visibility control	Yes	Yes	Yes	Yes	Yes
Blocking service requests	Yes	Yes	Yes	Yes	Yes
Subscription authorization	Yes	Cracked	Yes	Yes	Yes
Revoke authorization	Yes	No	No	No	No
Security					
Authentication encryption	Hybrid	No	Symmetric	Symmetric	Symmetric
Data encryption	Yes	No	No	No	No

OPNS, Online Presence Notification System; ICQ, I Seek You; AOL, America online; TCP, Transmission control protocol.

3.5 INTERNET TELEPHONY

Internet telephony systems are basically synchronous distributed systems whereby two users who are physically separated are able to carry out real-time voice communication over the Internet. Motivated by the huge potential cost savings by making transcontinental telephone calls at the prices of local telephone calls, plus nominal standard Internet connectivity charges, a wide range of commercial Internet telephony systems such as NetMeeting [30], Internet Phone [31] and VoxPhone [32] have been developed. With technological advancements in compression, buffering, dynamic rate control and networking, these systems have generally achieved reasonably good QoS of late.

However, two potential problems have emerged in the development of Internet telephony systems. First, the operating platforms that are currently supported by the various Internet telephony systems are predominantly for Windows 95/98 and Windows NT, while relatively few systems are developed for the Unix and Mac platforms. Second, Internet telephony systems are launched as standalone applications and must be downloaded and installed prior to operation. This means that when a user moves between different computers, new copies of the telephony software must be re-installed and configured before the actual communications can take place. Moreover, new versions of the software must either be purchased or downloaded before the existing software can be upgraded. This "stand-alone" nature of application-based telephony software causes much inconvenience for users.

To support a platform-independent Internet telephony environment, one possible solution is to develop Internet telephony systems using Java applet [33]. The use of Java applet removes the need to install the telephony system every time a user wants to talk to someone over the Internet. The user only needs to have a standard web browser with Java support (e.g. Netscape's Navigator and Microsoft's Internet Explorer) and a network connection. The necessary call set-up and voice communication procedures will be handled by the Java telephony applets and appear transparent to the end-users. This is analogous to making a telephone call without having to worry about the underlying workings of a telephony system.

3.5.1 Overview of an Internet Telephony System

Figure 3.8 shows the basic components of an Internet telephony system. Two host computers are required to serve as caller and recipient, where each host computer is identified by a unique IP address. Two modes of connection to the Internet are possible. Users can connect to the Internet either directly or via an Internet Service Provider (ISP). The host computer can either be a workstation or a PC with sufficient computation power and audio capabilities. The telephony system that resides on each host computer facilitates the real-time voice communication across the Internet.

In the basic communication process, the caller's telephony system acquires the real-time voice data through an audio input device and digitizes the analogue signals. The data is then compressed (and optionally encrypted) before being transmitted to the recipient through the Internet. Compression is necessary to reduce the bandwidth requirement of the data. At the recipient's end, the system carries out the reverse process. Incoming data is (decrypted), decompressed and played back on the audio output device of the recipient's computer. Communication can either be half or full duplex although full duplex is preferable since it emulates the conventional telephone system.

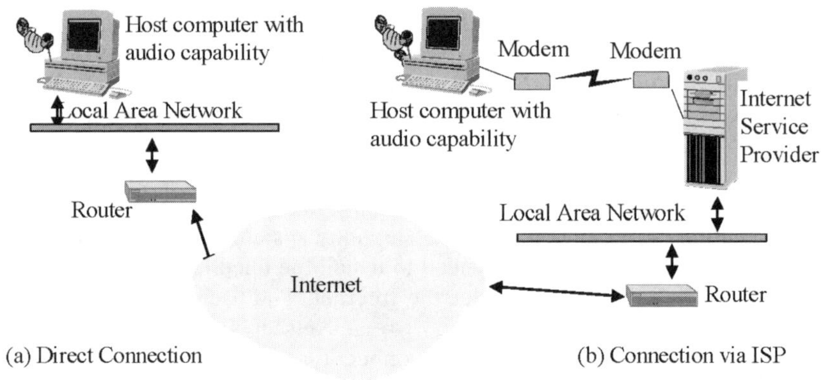

Figure 3.8 Overview of an Internet telephony system

As discussed in Section 3.1, the current best-effort Internet is not well suited to the delivery of real-time data, such as voice data for telephony applications. Different mechanisms have been developed to handle the delay jitters (timing uncertainties) and packet loss problems. The playout time of arriving audio packets can be adjusted at the destination to minimize the impact of delay jitters by using a buffering mechanism [34]. Various voice recovery methods such as silence substitution, waveform substitution, sample interpolation, embedded speech coding [35], Xor mechanism [36] and combined rate and control mechanism [37] have been proposed to minimize the impact of packet loss. With these mechanisms, most telephony systems have achieved acceptable quality of voice communications.

Chin *et al.* have applied a quality-based dynamic voice recovery mechanism [38] to enhance the quality and reliability of real-time voice communication. The dynamic recovery mechanism integrates the dynamic transmission control [39] with voice recovery [35, 37] using a quality-based measurement. It minimizes packet loss by controlling the transmission rate dynamically from the source based on the network congestion condition with a quality-based voice recovery at the destination. The quality of voice signals delivered is measured based on the subjective ratings of different voice codecs. Multiple redundancies are used to enable better reception and recovery of voice signals during congested network conditions.

3.5.2 Using Java for Platform Independence

Recently, the Java language has gained tremendous recognition and momentum. Its promise of "Write Once, Run Anywhere" platform independence means that source code only needs to be written and compiled once, and can then be distributed to different platforms. However, as Java is an interpreted language, programs written in Java often run slower than those written in native C or C++ code. This difference in performance is diminishing as the processor speed increases and the

Java language matures. This means that Java can now be used to perform computationally intensive tasks such as real-time Internet telephony.

As Java is still a relatively immature language, support in certain areas is lacking, especially in the area of multimedia. For several years since its inception, the multimedia capabilities of Java were only limited to playing a simple audio clip or performing simple image-based animation in a Java applet. There was and still is little support for audio and video capture and playback in Java. Due to this reason and various performance issues of Java [40], most telephony applications have been developed in other languages such as C or C++. Early applications have achieved this functionality in Java, but not without having to resort to accessing native audio methods, for example [40],. Recent technologies such as the Java Media Framework (JMF) [41] aim to give more multimedia support to Java, but they are still immature and unstable.

Java applets allow program code to be embedded into web pages and be downloaded "on-the-fly" through a standard web browser. This frees the user from having to install the telephony software each time he or she wants to talk with someone over the Internet. In this way, Java applets offer convenience and ease of use to the end-user. However, this can also be used by hackers to transmit viruses or other malicious code to the remote client systems. Therefore, certain security restrictions are placed to restrict what applets can do on the client system. Some of these restrictions include the opening of new network connections and accessing local files.

Sections 3.5.3 and 3.5.4 describe a web-based Internet Java Phone known as IJPhone. IJPhone is designed to be downloadable from the Internet and can be run from standard Java-compliant web browsers. The security restrictions that must be overcome in order to develop IJPhone to give end-users the convenience and ease-of-use of Java applets are presented in Chapter 4.

3.5.3 Internet Java Phone

The Internet Java Phone consists of two main components: Web-based Telephone Exchange (WTE) and Internet Java Phone (IJPhone) Applet. WTE provides a framework to allow multiple users to search and make calls to one another through a standard Java-compliant Web browser. The IJPhone Applet is a dynamically downloadable client software that provides the functionality required for real-time voice communication over the Internet.

Figure 3.9 illustrates the operations of the IJPhone system. The initiator of the call (the caller) and the receiver of the call (the recipient) firstly register with the WTE (step 1) through the Telephone Exchange Web Page. After logging in successfully, both the caller and recipient proceed to download the IJPhone Applet to their respective terminals and run the applet program (step 2). A list of users that are currently online is displayed on the Internet Java Phone Web Page, a web page generated by the WTE. The caller then selects the recipient's name from the list and sends it to the WTE to resolve it into the recipient's current IP address (step 3). The WTE resolves the recipient's name into its current IP address and returns it to the caller (step 4). The caller's IJPhone Applet takes the IP address and proceeds to establish a voice communication link to the recipient's IJPhone Applet (step 5). IP address resolution, IJPhone Applet-to-Applet call set-up and voice communication are transparent to the user.

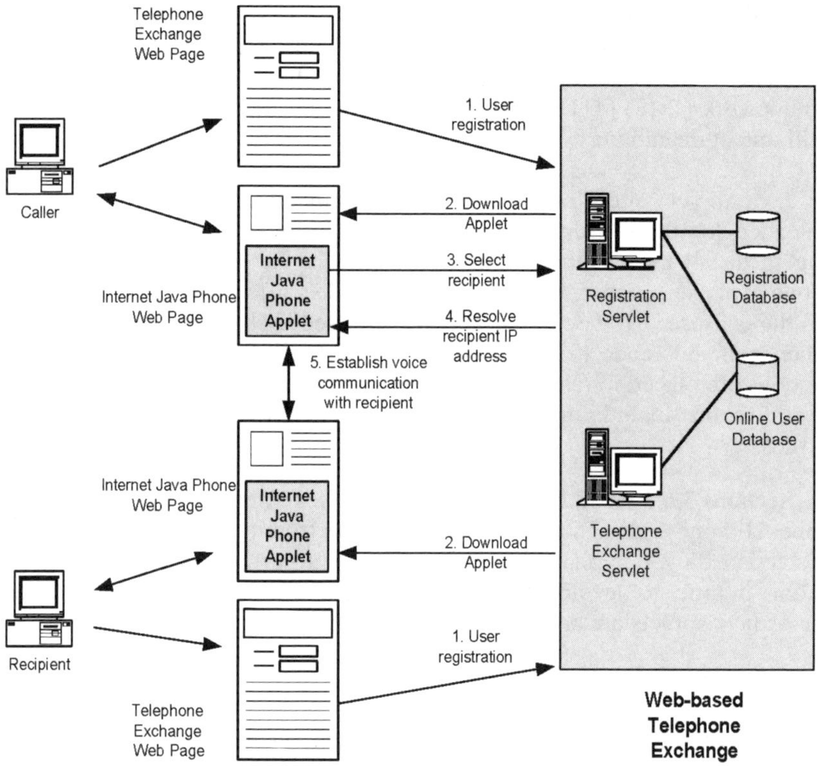

Figure 3.9 Internet Java phone system

3.5.4 Performance Comparison

The performance of the IJPhone system has been compared with popular Internet telephony systems including Microsoft's NetMeeting 3.01 [30], VocalTec's Internet Phone Release 5 [31] and VoxPhone 3.0 [32]. It should be noted that all these telephony systems were written in native code using C/C++. All the tests were conducted in a controlled and similar environment using the same equipment to ensure the consistency of the evaluation. The equipment used mainly includes two sets of Pentium II 350 MHz PC with 64 MB RAM running Microsoft Windows 95 and two sets of sound cards with full duplex audio drivers. The Java Servlet Development Kit (JSDK) 2.1 Web Server was used for the Web server. The criteria for comparison are the percentage of the central processing unit (CPU) usage, connection methods and downloading time. The comparison results are summarized in Table 3.3.

Table 3.3 Comparison of IJPhone with other Internet telephony systems

Telephony system	**Overall CPU usage**	**Connection methods**	**Downloading time**		
			File size (kb)	**56 kbps modem**	**10 mbps LAN**
IJPhone	20%	Web-based User List; Easy to connect	998	2.4 m	0.62 s
NetMeeting	8%	Internet Locator Servers (ILS); Difficult to connect	1,596	3.9 m	1.0 s
Internet Phone	18%	Community Browser; Moderate to connect	8,572	20.9 m	5.4 s
VoxPhone	10%	Online User List; Easy to connect	1,794	4.4 m	1.1 s

CPU, central processing unit; LAN, local area network.

3.5.4.1 Central Processing Unit (CPU) Usage

The overall CPU usage is measured by executing the same telephony software on the two different terminals. A voice connection is initiated between the two terminals and voice communication then takes place. The System Monitor (a system performance tool in Windows 95) is then invoked in the background to monitor the CPU usage by the telephony software over one minute; making records over five second intervals. The average of these values is then calculated to give the overall CPU usage.

Table 3.3 shows that IJPhone had the highest overall CPU usage. This is due to the fact that Java is an interpreted language and must be run on top of an interpreter like the Java Virtual Machine. However, the CPU usage of IJPhone is only 11% more than Internet Phone and 2.5 times that of NetMeeting. Although the performance of the IJPhone is slower than that of the other telephony systems coded in C/C++, it still performs well enough to support real-time Internet telephony. Also, with the increase of processor speeds and the maturing of the

Java language, the difference between the performance of Java and C/C++ based applications should fall in time.

3.5.4.2 Connection Method

All systems surveyed make use of some sort of telephony server implementation to allow users to view the presence of other online users and initiate calls to them. NetMeeting makes use of Internet Locator Servers (ILS) to register the IP addresses of users and allow searches to be conducted on online users. Internet Phone makes use of the Community Browser, which employs the concept of Internet Relay Chat (IRC) servers, having multiple servers linked together. Once a user is logged in, he or she can choose to participate in one of many "channels" to view other online users. VoxPhone makes use of a simple Online User List similar to that of IJPhone. However, the online user lists of these systems can only be accessed through the client software itself. This means that users must have the software pre-installed before seeing who is online. IJPhone has the added feature of allowing the online user list to be viewed on a Web site using a standard Web browser. From this, users can then decide whether to place a call before starting the telephony software.

The ILS used by NetMeeting has been found to be frequently congested and difficult to log into. Internet Phone separates its user directories over multiple servers; they are relatively easy to log into even when there are a large number of users already logged in. The IJPhone and VoxPhone servers are relatively easy to log into, but this is probably due to the small number of users as compared to those of NetMeeting and Internet Phone.

All four systems return a list of online users to allow users to view and pick another user to call. Internet Phone provides updates on user information periodically and reflects the change resulting from users logging on and off from the system. This makes the information more accurate. IJPhone downloads the online user list after being downloaded and run from the IJPhone Web page. Any subsequent change to the list is not reflected automatically and the user must manually request an updated list from the Telephone Exchange. However, this can be improved by sending update requests to the Telephone Exchange Servlet to update the online user list automatically.

3.5.4.3 Downloading Speed

As the IJPhone system is a web-based telephony system, it does not require any pre-installation before the system is being used. All the other telephony systems require the software to be pre-installed on the local machine, and this can be inconvenient and time-consuming for the user. This is especially true if the user does not have a fixed terminal or must travel around frequently. With IJPhone, the users can access it via any Java-compliant Web browser.

Table 3.3 also shows a comparison of the time taken to download the four different telephony systems. Two configurations are used to measure the time taken to download each of the telephony systems over the Internet — one using a 56 kbps modem, and the other using a 10 mbps LAN connection. Internet Phone has the largest file size and hence takes the longest time to download. The file size of the IJPhone is slightly under 1 MB and can be downloaded under three minutes in favourable network conditions using a 56 kbps modem. The file size must be kept small to minimize the waiting time for users of the IJPhone system.

3.6 VIDEO DATA TRANSMISSION

Transmission of real-time video data poses many problems as there are severe constraints on bandwidth resources, packet loss and delay that affect the perceived quality of video playback. A fundamental difference between text data and real-time video data is that, packet loss is intolerable for text data, while delay is acceptable; for real-time video data, some visual quality degradation is often acceptable while delay must be strictly bounded. This feature introduces many new error control mechanisms applicable to video application, but not applicable to traditional data.

In addition, the heterogeneity and current best-effort nature of the Internet offer no QoS reassurances. Figure 3.10 depicts a point-to-multipoint video transmission (multicast) system, which highlights the problems due to network heterogeneity and receiver heterogeneity.

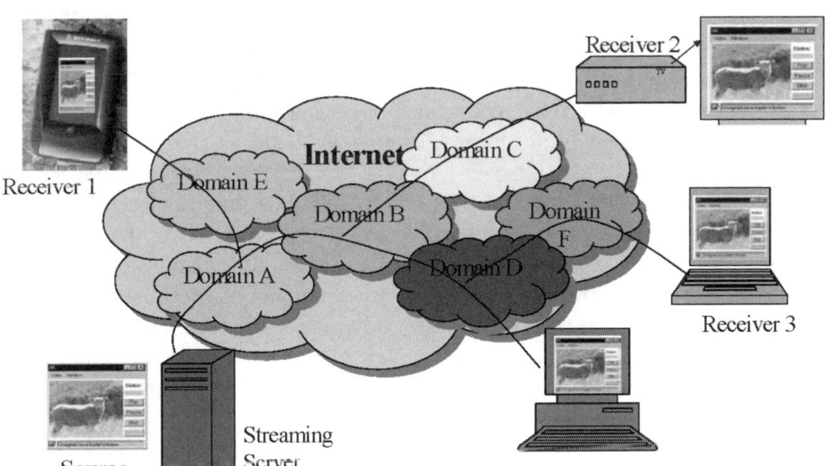

Figure 3.10 Point-to-multipoint video transmission

The fundamental problems associated with video over the Internet are therefore:

- *Bandwidth constraints* — real-time video transmission requires an absolute minimum bandwidth of 28 kbps to ensure adequate playout quality. The current Internet based on IPv4 does not provide such guarantees that the available bandwidth would be maintained at a certain rate throughout the duration of a video session. Excessive traffic can cause congestion, further degrading throughput performance. Although some newer routers provide congestion control, most do not.
- *Delay* — real-time video data requires strictly bounded end-to-end delay. In particular, every packet must arrive in time to be decoded/displayed to ensure continuous playout. Consequently, a packet that arrives late at the receiver will be discarded and this contributes to the total packet loss rate. Again, the current best-effort Internet does not provide any delay guarantees. Further, congestion can cause varying amounts of delay.
- *Packet loss* — the effects of packet loss range from disturbing to catastrophic, depending on the nature and quantity of packets lost. Current IP offers no packet loss guarantee. Loss can be high during peak network usage due to congestion.
- *Network heterogeneity* — sub-networks that make up the Internet have unevenly distributed resources (processing capabilities, bandwidth, storage, congestion control, and so on). Consequently, different users can experience different loss/delay characteristics, which may also be time varying.
- *Receiver heterogeneity* — receivers can have very different latency requirements, processing power, storage, display capabilities, and so on. All these may also vary with time. For example, some receivers may be handheld devices with limited processing power/memory, and so on, while others may be desktop workstations. Also, requirements may be application-dependent. For example, in videoconferencing, it may be desirable to impose more stringent real-time constraints on the active speaker, while passive listeners may prefer to trade latency for better quality.

The challenges of sending real-time video over the Internet have created much research interest. One major thrust of research is in improving the QoS for Internet delivery of streamed video, which refers to real-time transmission of buffered video [42]. Video streaming is preferred to downloading because the latter mode of delivery requires the whole file to be downloaded before video playback can begin. Streamed video, in contrast, allows playback to begin as soon as some of the video sequences have been received and decoded without having to wait for the entire file to arrive.

3.6.1 Video Streaming

As shown in Figure 3.11, a video streaming system comprises of three major components: a streaming server (sender), the Internet and a client (receiver). The streaming server processes video data prior to transmission. It also supports VCR-like functions such as fast-forward, rewind, play/pause, and so on. Raw video data must be suitably compressed before transmission. For non-scalable video encoding, video data is first transformed, typically using discrete cosine transform (DCT). This transform achieves the dual goal of energy compaction and signal decorrelation. This is then followed by quantization, run-length coding and entropy coding. Thus, non-scalable video compression results in a single bitstream. This approach is not optimal for video transmission over the Internet. Scalable video compression, in contrast, allows graceful degradation of video playback quality. This is often achieved by compressing raw video data into a base layer bitstream, together with one or more bitstream(s) for enhancement layer(s) [43]. Scalability is achievable both for DCT and wavelet transform [44]. The compressed data is then stored in the buffer and processed by an application layer QoS control stage before transmission.

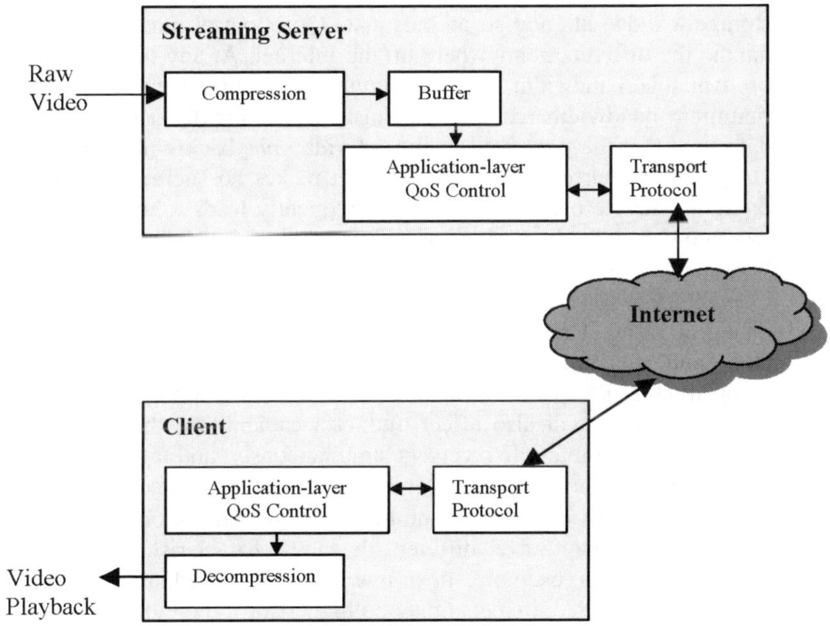

Figure 3.11 Video streaming system

The Internet serves as a continuously available channel for multimedia information transfer. While the Internet offers many benefits, such as worldwide coverage, there are many issues that have to be addressed to make it viable for video streaming applications. By adopting the Internet as a communication channel, we have to consider three types of protocols: network-layer protocol, transport protocol and application protocol. The nature of the Internet means that application layer QoS control plays a vital role in ensuring that the quality of video data delivered across the Internet is acceptable at the receiver.

Finally, the client basically performs the reverse operations of the server. For Web-based applications, video playback is performed via a Web browser.

3.6.2 Quality of Service Issues

Although the Internet offers a convenient medium for multimedia distribution, many issues have to be resolved, especially for transmission of real-time video. In particular, the following QoS issues must be taken into consideration.

- *Network Conditions* — the Internet is a heterogeneous environment that handles different types of data traffic. Since different sub-networks of the Internet have different resource management strategies with respect to storage, bandwidth, and so on, adequate QoS control must be in place to handle the differences anywhere on the Internet. At any time, the amount of available bandwidth may be limited and unpredictable. Usually, a minimum bandwidth requirement must be met for the delivery of video data such that the perceived quality of video playback will be acceptable. However, the current best-effort Internet makes no such guarantees. Furthermore, severe network congestion frequently leads to excessive packet loss and delays, both of which are detrimental to the quality of video playback.
- *Computing Resources* — sufficient computing power is required for video data processing. This includes scalable video compression, video transmission and reception, as well as constructing parity packets and reconstructing lost packets whenever necessary. The amount of processing power available will also affect multicast capabilities: the ability of the sender to handle multiple receivers simultaneously; and the ability of a receiver to receive and process several video streams at a time.
- *Receiver Heterogeneity* — in a multicast video distribution system, different receivers may have different latency characteristics and quality requirements. For example, in a live security surveillance session, the playback quality requirements may be higher for a monitored site that has suspicious activity occurring, rather than others.

The above issues sometimes cause contradictory requirements, as receivers may have different requirements from what the network can provide. Techniques that resolve these problems fall into two broad categories: network-based and sender/receiver-based. In the first category, strict QoS requirements are imposed on network routers to provide guarantees on bandwidth, delays, packet loss, and so on, typically by providing some form of resource reservation [45–49].

Techniques that fall under the second category make no such QoS assumptions on the network. These techniques are preferable, as they lead to more robust solutions and are compatible even with very bad network conditions. The heterogeneous nature of the Internet, compounded by the varying rates of acceptance of the new IPv6 means that application layer QoS control will continue to play a vital role in multimedia transmission for the foreseeable future. Thus, the remainder of this section focuses on sender/receiver-based techniques, which rely heavily on application layer QoS control mechanisms, for addressing the issues discussed above.

3.6.3 Application layer Quality of service Control

Application layer QoS control techniques may be viewed from two perspectives: transport and compression [50]. Control techniques that are viewed from a transport perspective are employed without reference to any particular video coding scheme. Thus, these techniques are semantically decoupled. Compression perspective is the opposite, where control techniques are applied to video semantics within a compression framework. In addition, application layer QoS control entails two operations: congestion control and error control.

3.6.3.1 Congestion Control

Congestion control is employed to reduce the impact of packet loss and delays caused by network congestion [51]. There are currently three major techniques for congestion control: rate control (transport perspective) [52], rate shaping (transport and compression perspectives) [53], and rate-adaptive video encoding [54] (compression perspective).

- *Rate Control* — UDP is typically used for transporting real-time video. Since UDP does not provide QoS guarantees, it is necessary to employ rate control at a higher layer (i.e. application layer). Rate control reduces network congestion by matching transmission rates with available bandwidth at any given time. In the absence of rate control, the packets transmitted at a rate higher than what could be supported by the prevailing network bandwidth would be lost. Rate control may be employed at the sender (source-based), receiver (receiver-based), or both (hybrid). In any case, there are two major approaches for determining network conditions: probing and modelling. In the probing approach, the sender and/or receiver implicitly estimate the network condition by conducting probing experiments for the available bandwidth [55]. In the model-based approach, the sender and/or receiver explicitly determine the network condition by using a throughput model based on a TCP connection [51].

- *Rate Shaping* — once the network condition is known, rate shaping (or filtering) is used to constrain the data rate to match the available bandwidth. Rate shaping may be considered from a transport perspective [56] or compression perspective [57]. There are currently five major types of filters employed for rate shaping [42]: codec filter, frame-dropping filter, layer-dropping filter, frequency filter and re-quantization filter.
- *Rate-adaptive Video Encoding* — adaptive encoding is employed in video coding standards, such as MPEG-1/2 [58], MPEG-4 [59] and H.261/263 [60], to improve the subjective quality of video playback under prevailing network conditions. Adaptation is performed by altering the quantization parameter (QP), video frame rate and/or video object rate. Fundamentally, all of these adaptation approaches operate within the framework of (rate-distortion) R-D theory [61].

3.6.3.2 Error Control

Error control is necessary to reduce the inevitable impact of bandwidth constraints, packet loss and delays on the quality of the received video. In the absence of error control, it is virtually impossible to provide an acceptable quality of playback with real-time video data delivered across the Internet. There are currently four major approaches used for error control: retransmission, forward error correction (FEC), error resilient coding and error concealment.

- *Retransmission* — considered from a transport perspective, retransmission can provide good error rates without incurring much bandwidth overhead because packets are retransmitted only when there are some indications that they are lost. However, retransmission may lead to intolerable delays for real-time video applications. Current solutions to constrain the retransmission delay revolve around adding extended control mechanisms to ensure that retransmitted packets arrive in time [62].
- *FEC* — with FEC [63], redundant information is transmitted along with the original information. Thus, lost packets can be recovered directly from the redundant information at the receiver. The idea is that retransmission incurs additional latency due to retransmission of missing packets. FEC (forward error correction) does not suffer from this drawback. FEC mechanism is more effective when lost packets are dispersed throughout the stream of packets sent from a source to a destination [64]. Therefore, it is well suited to real-time multimedia applications over the Internet. FEC may be implemented in channel coding (transport perspective) [65], source coding (compression perspective) [66] and joint source/channel coding (transport and compression perspectives) [50]. The channel coding approach is often preferred as it could be applied to generic data without

making references to any specific compression semantic. A simple FEC scheme that uses the exclusive-OR function is proposed in [67], in which a redundant packet is generated by exclusive-ORing every k[th] packet with the other *k* packets. This is then transmitted to the receiver along with the other *k* packets. However, this mechanism increases the source transmission rate by a factor of 1/*k*. Also, additional latency is introduced since *k* packets have to be received before the loss packet can be recovered.

- *Error Resilient Coding* — this is traditionally employed at the transmitter to reduce synchronization uncertainties in video transmission over wireless channels, for example, [68]. Recently, error resilient techniques have also been applied to Internet video transmission [69]. However, since resilience techniques are developed from a compression perspective, it is necessary to apply them to specific video coding semantics.
- *Error Concealment* — this is implemented at the receiver to reduce the visual impact of data lost during transmission, since data loss cannot be completely eradicated, but rather, it can only be minimized. The general approach is to either perform interpolation spatially or temporally to recover lost data [70]. Like error resilient coding, error concealment techniques are considered from a compression perspective.

Figure 3.12 summarizes the QoS control techniques described above. Currently, FEC offers the best potential for real-time Internet video transmission. In addition, it is possible to combine FEC with error resilience coding and/or error concealment, for example, [71].

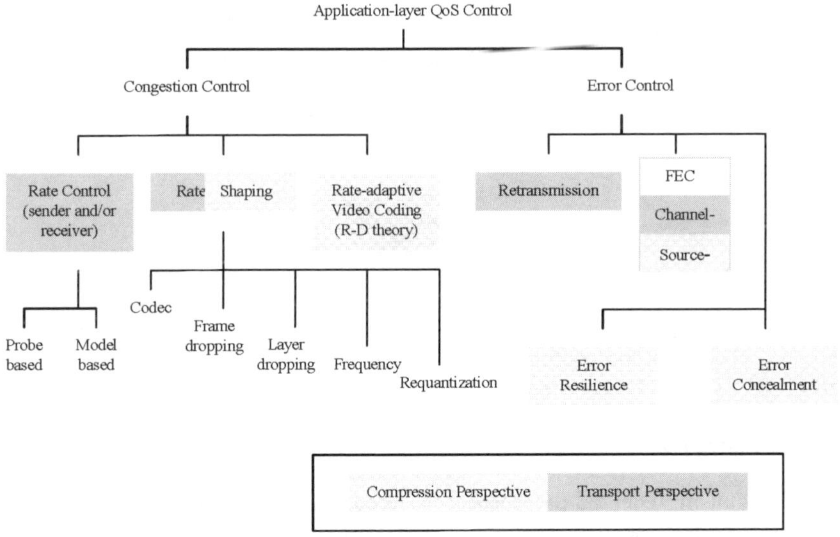

Figure 3.12 Application layer QoS Control mechanisms

3.6.4 Adaptive Transmission and Recovery Mechanism

Based on the above discussion, the authors have developed a unified solution that deals with both congestion control and error control for real-time multicast of streamed video. As shown in Figure 3.13, the adaptive transmission and recovery mechanism (ATRM) involves four stages: Packet Loss Analysis, Network State Estimation and Bandwidth Adjustment (Rate Control), Video Data Stream Determination (rate shaping) and Adaptive Error Control. This mechanism can effectively adapt to network conditions. In fact, ATRM and a slightly modified version known as integrated QoS control mechanism (IQCM) [72] have been successfully applied to the transmission of video data for security surveillance [73] and educational purposes [74].

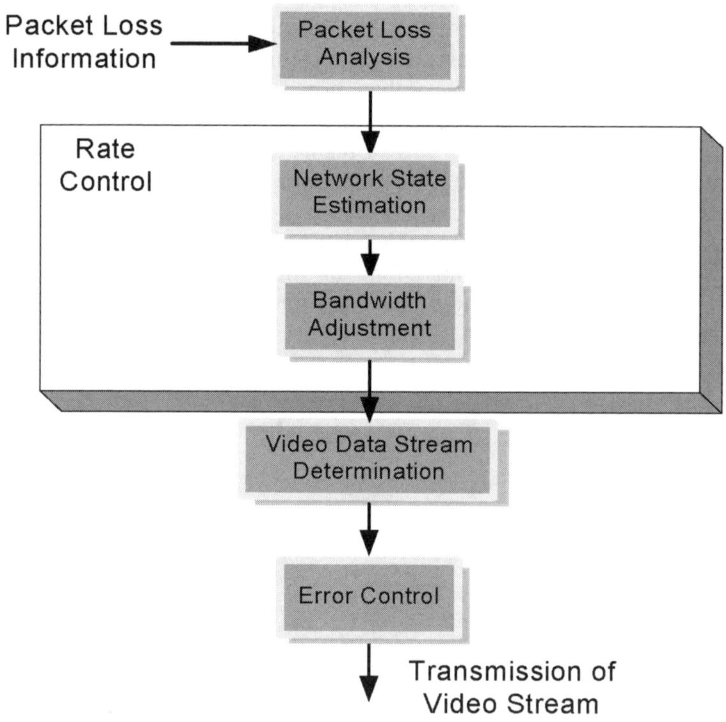

Figure 3.13 The four stages of ATRM

3.6.4.1 Packet Loss Analysis

In this mechanism, received packets are buffered at the receiver. A packet loss is detected using the RTP sequence number [8]. In particular, packet *n* is considered lost if it is still missing after packets *n*+1, *n*+2, and *n*+3 have arrived. Based on the loss statistics reported, the sender determines the loss rate for each receiver. A low-pass filter is used to smooth out the loss rate. The new smoothed loss rate λ_{new} is computed according to (3.1).

$$\lambda_{new} = (1-\alpha)\,\lambda_{old} + \alpha b \tag{3.1}$$

where

b	new loss rate.
α	constant between 0 and 1 used to indicate the influence of the new loss rate on the final smoothed loss rate.
λ_{old}	old smoothed loss rate.

A moderate value of 0.3 for α has been used as discussed in [39].

3.6.4.2 Rate Control

In the case of a point-to-point connection (unicast), the network congestion state can be mapped directly to decrease, hold or increase the current rate. In a multicast system, since several receivers may connect to the sender simultaneously, the sender maintains a point-to-multipoint connection. Thus, a model-based rate control mechanism is implemented at the sender. This is performed by Network State Estimation and Bandwidth (or rate) Adjustment. The smoothed value of the loss rates from stage one is used as a measure of network congestion. The individual network congestion state is determined and used to make the decision of increasing, holding or decreasing the bandwidth of the connection for each receiver.

As shown in Figure 3.14, two predefined thresholds, λ_c and λ_u are used to determine the network state seen by each receiver as UNLOADED, LOADED or CONGESTED. They are defined according to the users' perception toward different playout qualities of received video data due to varying degrees of packet losses. The upper threshold λ_c is defined as the upper limit, such that the received video quality is unacceptable if this threshold is exceeded. The lower threshold λ_u is the lower limit, such that packet loss rates below this limit will give good video playout quality. For loss rates between these two thresholds, the quality of received video data is considered acceptable to users.

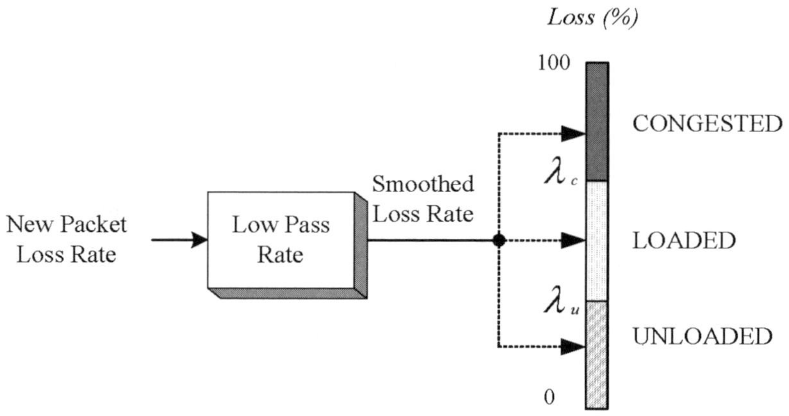

Figure 3.14 Classification of network congestion state

The values for λ_c and λ_u have been determined experimentally. In the experiments, a loss rate higher than 25% gave unacceptable perceived quality of video playback, whereas video with less than 10% of losses was considered to be of good quality. Consequently, $\lambda_c = 0.21$ and $\lambda_u = 0.18$ were chosen for the mechanism.

Once Network State Estimation is complete, the source performs Bandwidth Adjustment for each receiver individually, according to the estimated network congestion state. The sender must rapidly reduce the bandwidth in case of congestion. A multiplicative decrease σ is used for the CONGESTED state and an additive increase ρ is used for the UNLOADED state. If the network state is LOADED, no changes take place. Similar linear increase/multiplicative decrease algorithms have been discussed in [39]. It is also important to ensure that the bandwidth is always larger than some minimum bandwidth, $bandwidth_{min}$, to guarantee a minimum quality of the video playout at the receivers. Similarly, a maximum bandwidth, $bandwidth_{max}$, is also defined as shown in Figure 3.15.

```
if ( network_state == CONGESTED )
        bandwidth_allow = max (bandwidth_old * σ, bandwidth_min);
else if ( network_state == UNLOADED )
        bandwidth_allow = min (bandwidth_old + ρ, bandwidth_max);
```

Figure 3.15 Bandwidth adjustment algorithm

In this algorithm, the value $bandwidth_{old}$ is the actual bandwidth adjusted for each receiver in the most recent control actions taken by the sender. The value

$bandwidth_{allow}$ is the currently allowed bandwidth that can be used by the sender for the same receiver.

3.6.4.3 Video Data Stream Determination

Rate shaping is required for the sender-based rate control in stage 2, since the pre-compressed data rate may not match the target network bandwidth [42]. In particular, the sender must provide scalable quality of video streams for different receivers according to their currently allowed bandwidth and their network congestion states. We have adopted a rate shaping technique that is similar to re-quantization filtering [75], where transform coefficients are de-quantized from the compressed data followed by a coarser re-quantization. As shown in Figure 3.16, our rate shaping mechanism does not require de-quantization and re-quantization. Instead, multiple video streams are produced using different QPs.

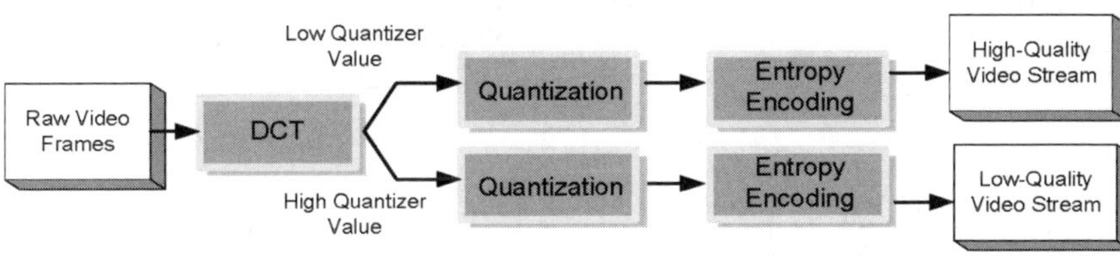

Figure 3.16 Rate shaping with different QP values

In a video session, the sender performs a two-level quality video coding and transmits the high-quality video stream to receivers with good network conditions, and the low-quality video stream to those with congested network conditions. The sender can dynamically switch between high-quality and low-quality video streams to account for changing network conditions. Figure 3.17 shows the video streams switching algorithm, which has a predefined bandwidth threshold specified as $bandwidth_0$. A buffer value $bandwidth_b$ is introduced to avoid highly responsive switching actions.

```
if ((current_stream = = low_quality_stream) &&
      (network_state = = UNLOADED) &&
      (bandwidth_new > bandwidth_0 + bandwidth_b) )
   current_stream = high_quality_stream;
else if ((current_stream = = high_quality_stream) &&
      (network_state = = CONGESTED) &&
      (bandwidth_new < bandwidth_0 – bandwidth_b) )
   current_stream = low_quality_stream;
```

Figure 3.17 Algorithm for switching between video streams

3.6.4.4 Adaptive Error Control

The error control mechanism is based on the FEC scheme, which adapts to prevailing network conditions. Data packets are grouped into blocks of predetermined sizes, and a parity packet that contains the error control bits is inserted into each block. The block size for constructing parity redundant packets is determined according to network conditions. The receiver can then use the parity packets for the recovery of packets lost during transmission. A parity packet P is constructed using the exclusive-OR function as in (3.2).

$$P = D_0 \oplus D_1 \oplus D_2 \oplus \ldots \oplus D_{K-1} \qquad (3.2)$$

where

D_i $(i = 0, 1, \ldots.K-1)$ i-th packet.

For simplicity, all packets are assumed to be m bits long; however, the scheme will also work for variable length packets. To each block of K data packets, the source adds an m-bit parity packet, whose i-th bit is given by (3.3)

$$c_{K,i} = \left(\sum_{j=0}^{K-1} c_{j,i} \right) \bmod (2)\,,\ i = 0, 1, \ldots m-1, \qquad (3.3)$$

where

$c_{j,i}$ i-th bit of the j-th packet.

The block size, which may take one of three values 4, 2, and ∞, can be adjusted to adapt to the variable network conditions by the source. An infinite block size indicates that no redundant parity packet is needed in video transmission. The block size and the construction of the parity packets determine the ability of recovering lost data packets. The receiver can reconstruct any missing packet using the other $K-1$ data packets and the parity packet. Figures 3.18 (a) and (b) each show an example of generating a parity packet and recovering a lost packet. In each case, the block size is 4.

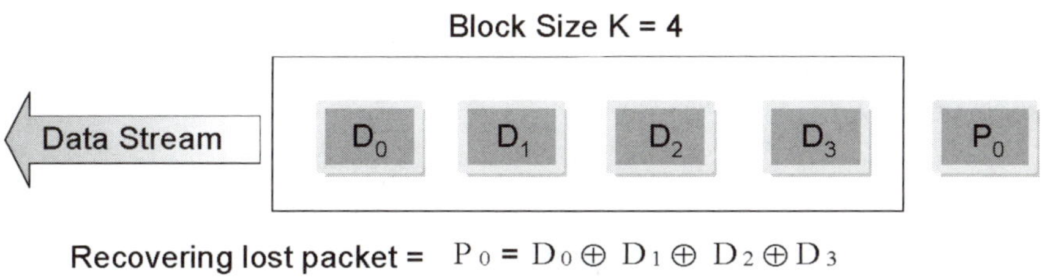

a) Generating a parity packet

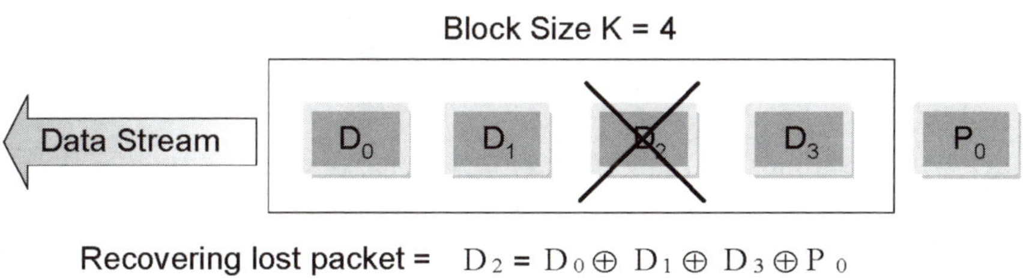

b) Recovery of a lost data packet

Figure 3.18 Parity packet, block size = 4

Since smoothed loss rate is computed based on accumulated past loss rates, it reflects a long-term reception condition. On the other hand, the current loss rate reflects a short-term reception condition. Also, a decreasing current loss rate will have the effect of reducing the smoothed loss rate. Therefore, the block size is determined based on the highest value of these two computed loss rates to increase the robustness of the mechanism. In order to improve the quality of the received video data, ATRM uses parity redundant packets to reduce the packet loss to within λ_u. When the loss rate is below λ_u, it indicates good video quality. Figure 3.19 shows the determination of the block size that is dependent on two values, the Upper Loss Limit (U) and Lower Loss Limit (L), which are functions of λ_u as in (3.4) and (3.5).

$$L = \lambda_u \tag{3.4}$$

$$(1-U)+U(1-U)=1-\lambda u$$
$$\Rightarrow \qquad U=\sqrt{\lambda u} \tag{3.5}$$

(6.4.4) defines the Lower Loss Limit where no parity packets are needed to constrain losses to λ_u. Above this loss limit, parity packets are needed to help recover from the higher losses and the block size is set to four. However, when the packet loss exceeds the Upper Loss Limit in (6.4.5), the block size is set to two to maintain losses to within λ_u.

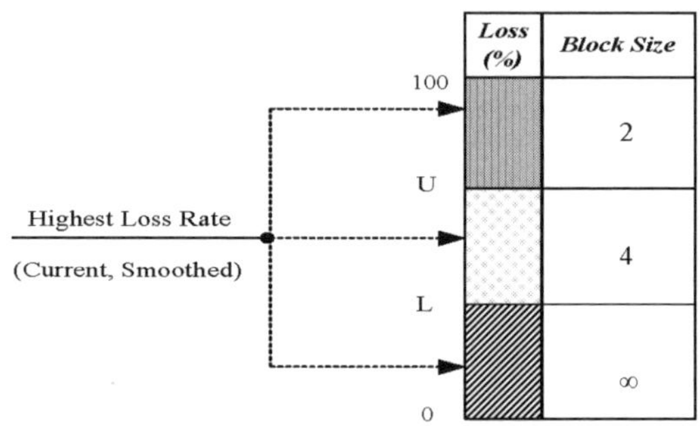

Figure 3.19 Determination of block size for video packets

3.6.4.5 Example of Adaptive Transmission and Recovery Mechanism (ATRM) Application

To illustrate how ATRM works, an experiment has been conducted using the values $\lambda_c = 0.21$, $\lambda_u = 0.08$, $\alpha = 0.3$, $\rho = 8.5$ kbps, $\sigma = 0.75$, so $U = \sqrt{\lambda u} = \sqrt{0.08} = 0.28$ and L = 0.08. A high-quality video stream and a low-quality video stream were generated by the source simultaneously. The values for $bandwidth_0$ and $bandwidth_b$ were found to be 51.0 kbps and 8.5 kbps, respectively. In addition, we set $bandwidth_{min}$ and $bandwidth_{max}$ to 10.6 kbps and 170.0 kbps, respectively. We connected three receivers to a sender for video transmission. Table 3.4 shows the status of each receiver at the start of the experiment.

Table 3.4 Status of receivers

Receiver #	Old smoothed loss rate (λ_{old})	Current loss rate (b)	Current bandwidth	Current video stream
1	0.22	0.25	45.0 kbps	High-quality
2	0.16	0.13	28.5 kbps	Low-quality
3	0.05	0.07	54.3 kbps	Low-quality

For each receiver, the new smoothed loss rate was calculated using (6.4.1), and the new network congestion state was estimated. Based on the bandwidth adjustment algorithm and the video streams switching algorithm, the new $bandwidth_{allow}$ and new video stream were determined and are shown in Table 3.5.

Table 3.5 Updated status of receivers

Receiver #	New smoothed loss rate (λ_{new})	Estimated network state	New allowed bandwidth	New video stream
1	0.230	CONGESTED	33.8 kbps	Low-quality
2	0.151	LOADED	28.5 kbps	Low-quality
3	0.056	UNLOADED	62.8 kbps	High-quality

Table 3.6 shows the determination of block sizes for constructing parity redundant packets for the receivers. For receivers one and two, since the highest values of the new smoothed loss rate and current loss rate were 0.250 and 0.151, which fell between Lower Loss Limit (L) and Highest Loss Limit (U), the block sizes (K) of both receivers were set to 4. On the other hand, the highest loss rate of Receiver three was 0.070 (video data were transmitted to this receiver.

Table 3.6 Block sizes for redundancy transmission

Receiver #	Max (b, λ_{new})	Block size (K)
1	0.250	4
2	0.151	4
3	0.070	∞

3.7 DESKTOP VIDEOCONFERENCING

Videoconferencing differs from video broadcasting/streaming in that the latter is a one-way process, while the former is a two-way process much like a traditional telephone conversation. In particular, a caller initiates a videoconference call for another party to answer. From the users' perspective, the main difference between a traditional telephone conversation and videoconferencing is that the latter allows the callers to see each other and to optionally exchange data files.

Table 3.7 Main characteristics of room-based and desktop videoconferencing systems

	Room-based	**Desktop**
Installation	Dedicated set-up; in special rooms	Anywhere via personal computers
Communication medium	ISDN	Internet
Costs	Relatively high installation and running costs	Much lower upfront and ongoing costs
Operation and Control	Typically requires professional operator with centralized control	Do-it-yourself approach with decentralized control
Scheduling	In advance	As and when required

ISDN, Integrated services digital network.

There are two main forms of videoconferencing: traditional room-based and desktop systems [76]. The main characteristics of these systems are summarized in Table 3.7. The International Telecommunications Union (ITU) H.320 standard was originally developed for videoconferencing over the Integrated services digital network (ISDN), which is a network with fixed rate channels (p x 64 kbps, $p \in [1,32]$). This means the original H.320 is not designed for packet-switched networks such as the Internet. Further, the current best-effort Internet does not provide the same degree of QoS as ISDN. A number of attempts had been made to adapt H.320 for videoconferencing over the Internet. However, this is now unnecessary for desktop videoconferencing with the advent of new standards, such as H.323 and H.324.

3.7.1 The International Telecommunications Union (ITU) H.3xx Standards

3.7.1.1 H.320

This is the oldest of the H.3xx series of ITU standards. It was developed for videoconferencing over ISDN, with transmission rates in the range between 64 kbps and 2 mbps. It is commonly used for rates in the range 64 kbps to 128 kbps. However, when the rate is below 129 kbps, the video playout can appear discontinuous. A transmission bit rate of 382 kbps or above reduces this apparent discontinuity, but this is often achieved by combining three (or more) 128 kbps channels into one, which greatly increases the cost and complexity.

3.7.1.2 H.321

In a bid to reduce the complexity and costs associated with H.320 deployment, ITU has introduced the H.321 standard. H.321 is designed to operate on Asynchronous transfer mode (ATM) network, with transmission rates similar to H.320. The key difference between H.320 and H.321 is that the latter offers additional QoS, which improves both the time delay and playout quality.

3.7.1.3 H.323

H.323 has gained widespread acceptance since its introduction, and is perhaps the most popular of the H.3xx series standards. In view of its importance, a more thorough treatment of H.323 is presented here. Like others in the series, the H.323 is in fact a suit of standards covering various aspects of real-time multimedia communications over packet-switched networks, such as a LAN and the Internet. H.323 has become popular because it offers fairly good quality at low costs (due to the use of PC and the Internet). In addition, it is compatible with H.320 terminals via a suitable gateway.

Specifically, H.323 covers voice (with the mandatory audio codec G.711 and optional audio codecs G.722, G.723, G.728, G.729), video (with the mandatory video codec H.261 and optional video codec H.263), data sharing (T.120 and T.125 Annex A (multicast)), and call control (H.245). A comparison of the G.7xx codecs can be found in [77]. Figure 3.20 illustrates the H.323 protocol stack. It is clear that delay-intolerant applications (i.e. real-time audio and video compression/decompression and real-time audio/video control) sit above UDP, while call control and data sharing sit above TCP for ensuring utmost data integrity.

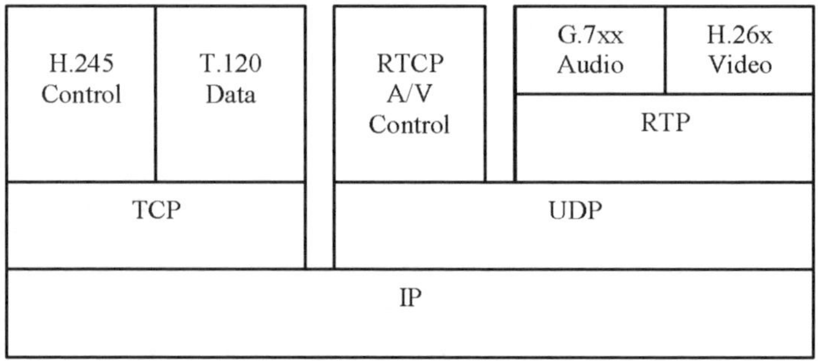

Figure 3.20 H.323 Protocol stack

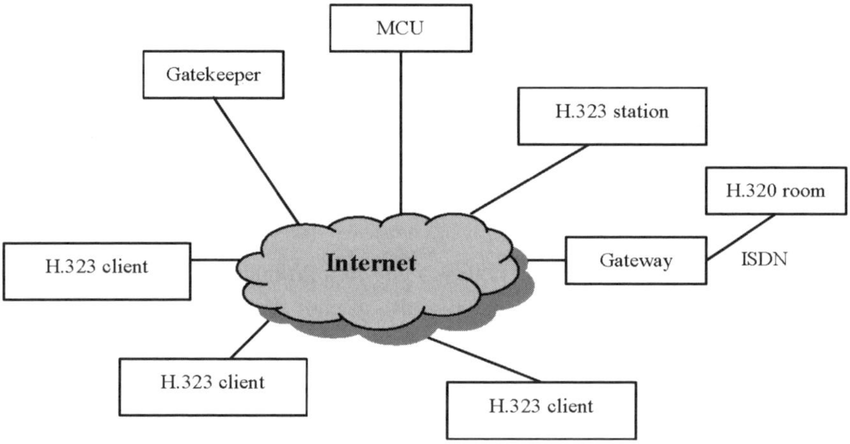

Figure 3.21 H.323 system set-up

Figure 3.21 illustrates the basic components that make up an H.323 system, which also shows an H.320 terminal connected via a gateway. The main components are

- *Gatekeeper* — this is the master control for all the other components in its zone. A zone includes a collection of H.323 devices that participate in a videoconferencing session. There can only be one active gatekeeper in any given zone. A gatekeeper's role is similar to a domain name server's in that it performs address translation, admissions control, bandwidth control and zone management. So, it provides calling with telephone numbers and aliases rather than IP addresses. Since a gatekeeper's role is one of control and management, videoconference calls do not actually go through a gatekeeper.

- *Multipoint control unit* (MCU) — also known as a "bridge", this is used for multipoint operations. More specifically, it acts as an Internet server to allow more than two people to be included in a videoconferencing session. It ensures that other participants always hear the voice of each speaker. Multiple MCU may be combined together in two ways: cascading for larger videoconferences involving more simultaneous participants; or used separately for more multiple videoconferencing sessions that occur in parallel. Currently, dedicated hardware MCU work well, whereas software implementations have lagged behind in performance.
- *Client* — this is the user's terminal, typically a PC.
- *Gateway* — this is optionally installed for communicating with non-H.323 terminals, such as H.320 terminals via ISDN, H.321 terminals over ATM, H.324 terminals over PSTN.

3.7.1.4 H.324

H.324 has been developed for multimedia communication on low-bit-rate circuit-switched networks, such as the PSTN, otherwise known as the *plain old telephone service* (POTS). In particular, H.324 has been designed to optimize low-bit-rate transmission performance in terms of video and voice quality. This has been made possible with the availability of V.34 data modems, as well as advancements in computing and compression technologies. With transmission bit rates limited to between 28.8 kbps and 64 kbps, it is not intended for applications that require high-quality playout. The primary appeal of the H.324 is therefore its low costs of deployment and operation.

3.7.1.5 H.310

Very good videoconferencing performances can be achieved with either H.321 or H.320 when the transmission rate is at least 784 kbps. However, H.310 offers even better quality due to the effective use of MPEG video compression, and significantly higher transmission bit rates between 8 mbps and 16 mbps.

3.7.1.6 Summary of H.3xx Standards

Table 3.8 provides a comparison of the various ITU H.3xx standards for multimedia conferencing.

Table 3.8 Summary of various H.3xx standards

Standard	Network	Bit rates	Video (basic)	Audio (basic)	Control	Relative quality (5 = best)
H.320	ISDN	64 kbps–2 mbps	H.261	G.711	H.242	3
H.321	ATM	384 kbps–2 mbps[a]	H.261	G.711	H.242	4
H.323	LAN-IP	128–384 kbps	H.261	G.711	H.245	2
H.324	PSTN	28.8–64 kbps	H.263	G.723.1	H.245	1
H.310	ATM	8–16 mbps	H.262/ MPEG	MPEG	H.245	5

ISDN, Integrated services digital network; ATM, Asynchronous transfer mode; LAN-IP, local area network-Internet protocol; PSTN, public switched telephone network; MPEG, Moving picture experts group.
[a]Although H.320 and H.321 often operate at the same transmission bit rates, the latter additionally includes QoS.

3.7.2 Session Initiation Protocol (SIP)

Similar to H.3xx protocol suites, Session Initiation Protocol (SIP) [78] is also a control-signaling protocol [77]. Significantly, SIP was developed specifically for IP networks. A fundamental difference between H.3xx and SIP is that while the former requires application access by means of call control, users of SIP can interact directly with applications without the need for call control.

Similar to HTTP, SIP can be used for the set up and subsequent tear down of multimedia communication sessions, including online presence and instant messaging. In particular, SIP uses HTTP's request-response model for its transactions. An SIP transaction begins with a request from a user to the server. This then activates a server function. Finally, the server responds to the request accordingly.

It uses a uniform resource identifier (URI) to identify a logical destination (nickname, e-mail address or telephone number) rather than an IP address. In fact, SIP defines logical entities based on a client-server model: when a user wants to initiate a session, he uses his terminal as a client to send a SIP request. The server then accepts the request and responds to it. In particular, six request methods have been defined [77]:

- *Register* — this provides a means for users to register their contact information with an SIP server.
- *Invite* — this initiates the session signalling sequence.
- *ACK* — this is an acknowledgement used to support session set-up.
- *Cancel* — this is used to abort a session set-up.
- *Bye* — this is used to end a session.
- *Options* — a major benefit of using SIP is that it can accommodate a wide range of SIP hosts having different degrees of sophistication and processing power. The options request allows a user to obtain information from a SIP server about its capability.

A major drawback of SIP is with regard to its security. Currently, encryption is the common approach adopted to counter security threats to the confidentiality of SIP requests and sensitive information contained about the users. However, it is difficult to ensure end-to-end security by encrypting all SIP requests and responses.

3.8 UNIFIED MESSAGING

The Internet has been used as a medium for many modes of person-to-person communication, such as e-mail, real-time voice, video, fax, paging, short messages, and so on, many of which have been discussed in earlier sections of this chapter. However, different equipment is often needed to access the various communication services. Further, the number and types of Internet-supported communication services are likely to increase with advancement in technology. How can a user manage all these communication services effectively?

In order to help users cope with the increasingly complex global information environment and the number of communication services, the concept of Unified Messaging [79] has been proposed for unifying various communication services into one single mechanism. This eliminates the need for multiple communication devices and the need for learning how to use them. The idea is to provide each user with an easy-to-use one-stop interface for accessing all types of communication services supported by the Internet.

Currently, a number of Unified Messaging Systems (UMS) such as JFax [80], 2bsure [81], MessagePoint [82] and iPost Universal Messaging [83] have been developed to support e-mail, fax and voicemail for users. These UMS use the concept of a Universal Inbox [84] for users to store and retrieve their messages. A universal inbox allows multiple media components and messages to converge upon a single point of retrieval for the recipient and providing a single point of message dispatch as well. UMS give each of their users a single phone number, which can be used to receive fax, voicemail and pager messages. This aims to merge the different numbers (e.g. fax number, voicemail number, pager number, mobile phone number) a user has for the different media into a single number. In these UMS, the e-mail system is typically used as the underlying transportation and messaging mechanism. Users are provided with an e-mail account, and the e-mail storage space then serves as the user's universal inbox. Therefore, UMS users can view their messages via e-mail client software or e-mail websites. In addition, users may also retrieve their messages via other devices apart from the standard equipment. For example, users may listen to their e-mails on the phone, or view fax messages with the e-mail software if they so wish.

3.8.1 Personal Communicator

The Personal Communicator [85] has been developed with the aim of integrating multiple personal communication services into a unified interface, and to provide a ubiquitous communication platform to users. It is designed to be a one-stop source that combines online presence notification with personal communication services such as e-mail, voicemail, facsimile, instant messagingi, text chat, voice chat and other real-time services. There are, in fact, two versions of the Personal Communicator: an application-based version for computing device owners to install on their desktop or notebook and a web-based version that allows access via web browsers using public terminals.

3.8.1.1 Application-based Personal Communicator

The user interface of the application-based Personal Communicator is shown in Figure 3.22. Users can view the list of online users that are in their subscription list and send instant messages to them or request to initiate real-time communication such as text chat or voice chat with them. Besides these services, users can also right click on the nickname of users in their subscription list to send e-mail to other users. Short cut buttons on the tool bar of the application also allow the messaging window to be launched for the checking of messages and the composition of new messages.

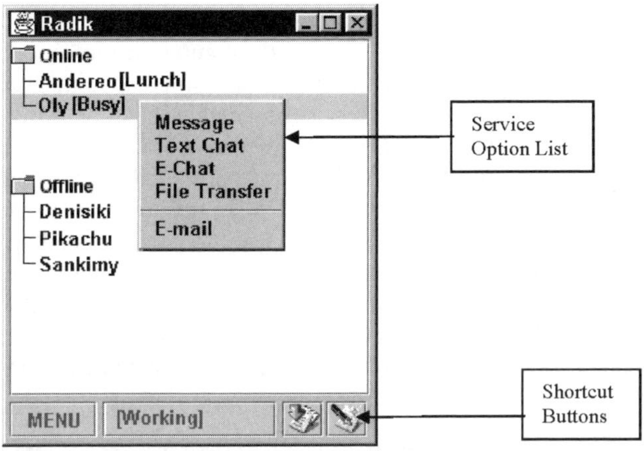

Figure 3.22 Application-based Personal Communicator

3.8.1.2 Web-based Personal Communicator

Figure 3.23 shows the message-checking interface of the Web-based Personal Communicator. Users are able to retrieve their e-mails, voicemails and faxes. They can also view a summary of the mail contents or preview any attachments. The online presence notification interface of the Web-based Personal Communicator is shown in Figure 3.24. Users can select the target user from the list of online users to send them instant messages or to have real-time chat with them.

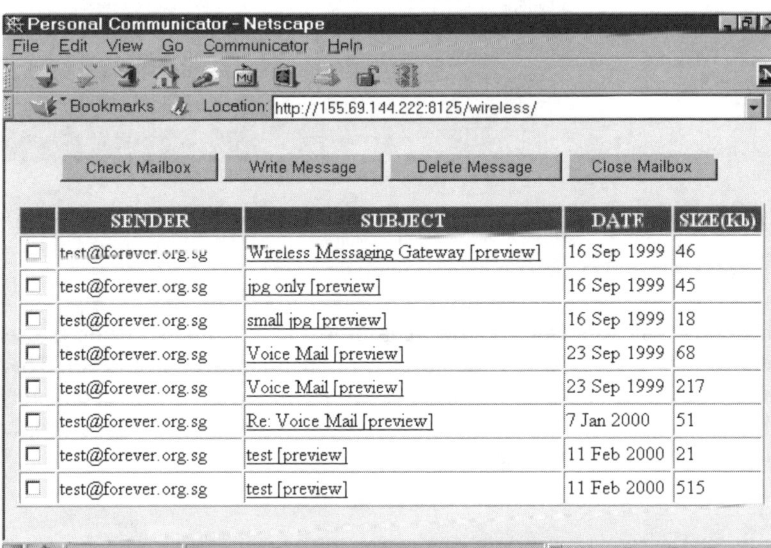

Figure 3.23 Web-based message-checking interface

Unlike the application-based Personal Communicator, the Web-based version does not allow users to add any third party communication modules. This is because the Web-based Personal Communicator's Web administrators add the communication modules. As such, the services available might be different from the application-based Personal Communicator. However, basic communication services such as instant messaging and text-based chat are provided. These communication services are powered by signed Java Applets. Unlike unsigned Java Applets, signed Java Applets are trusted and they have the privilege to perform reading and writing of data on the local disk and even make network connections with other computers beside the host computer that they have downloaded. Signed Java Applets are used because there is a need to make direct network connections when users communicate with each other. Making direct network connections between users will help to offload the server.

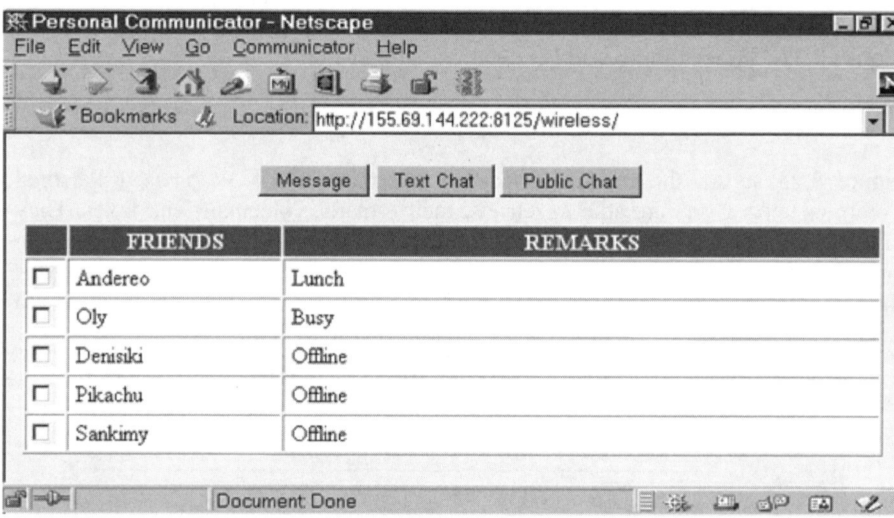

Figure 3.24 Web-based online presence notification interface

3.8.2 Real-Time Communication Services

Real-time communication services between users can be initiated using the service negotiation support provided by the online presence notification system. This section presents the open application interface for service negotiation support for real-time communication modules. This is followed by an implementation example of a communication module for instant messaging.

3.8.2.1 Open Application Interface for Service Negotiation

The online presence notification system mentioned in Section 3.4 does not provide any real-time communication services directly. However, its design allows others to provide real-time communication services by designing their communication application to use the online presence notification system's open application interface for service negotiation. In this way, the OPNS not only provides online presence information to users but also allows users to initiate real-time communication with each other. Figure 3.25 illustrates this concept.

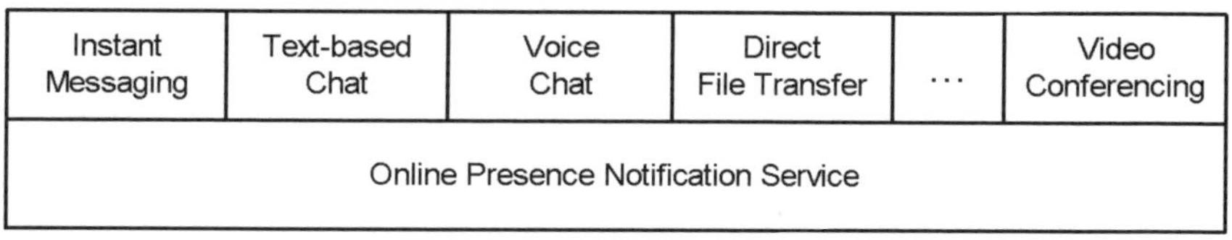

Figure 3.25 Real-time communication service support

Service negotiations are done via the online presence notification server, which relays communication requests only if both parties are in each other's subscription list and the requester is an unblocked authorized subscriber of the recipient's online presence information. The Personal Communicator will only process communication requests from the server, but not from other Personal Communicators or other applications. Therefore, users will not be harassed by communication requests from users that are not in their subscription list. In addition, the use of the hybrid encryption scheme also helps to ensure that communication requests are secure and authentic.

The Personal Communicator maintains a list of communication modules installed, and displays the different communication options when the nickname in the online presence list with a right click on the mouse. When a user requests for real-time communication services with another user, the requester's Personal Communicator will negotiate on behalf of the communication module as described in the Service Negotiation Process in Section 3.4. The communication module is not directly involved in the negotiation. The requester's Personal Communicator will inform the server to negotiate with the recipient's Personal Communicator.

If the recipient accepts the communication request, it will supply the server with a private session key and private session ID along with its IP address. The server will then relay them to the requester's Personal Communicator. The recipient's Personal Communicator will also launch the selected communication module with the requester's user ID and nickname, the requester's IP address, the private session key and the private session ID as parameters in the respective order. The launched communication module will then listen for incoming connection from the requester.

The requester's Personal Communicator will receive the recipient's IP address, the private session key and private session ID from the server if the recipient accepts the communication request. The requester's Personal Communicator will then launch the selected communication module with the recipient's user ID and nickname, the recipient's IP address, the private session key and the private session ID as parameters in the respective order. From then onwards, the launched communication module will interact directly with the equivalent communication module running at the recipient's side. Figure 3.26 illustrates the service negotiation process.

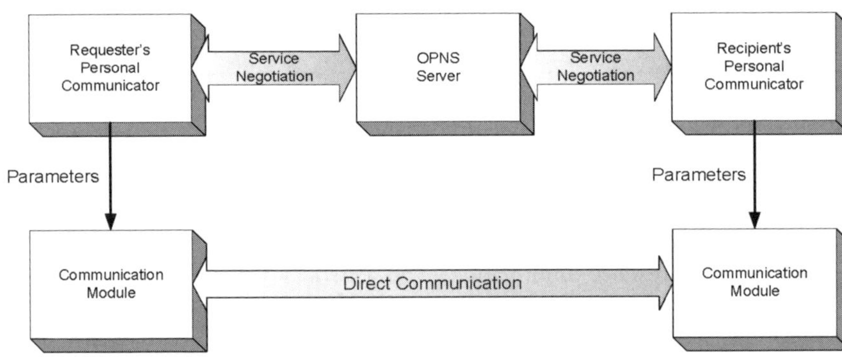

Figure 3.26 Service negotiation process

In cases where the recipient does not accept the communication request, the server will also relay the reply to the requester's Personal Communicator. In the event that the recipient did not install the same communication module or a compatible communication module that can provide equivalent service, an incompatible service request message will be returned to the requester.

In order to provide real-time communication services via the online presence notification system, the third party communication module developers need to make provisions for their communication modules to register with the Personal Communicator. This is done by supplying a special data file, *register.dat*, which contains the full path name of the communication module, a service name and a service option display name to the Personal Communicator. The communication module's installer can launch Personal Communicator's communication module manager to read this special file so as to register the module to the Personal Communicator's registry. Figure 3.27 shows the content of a sample register.dat file.

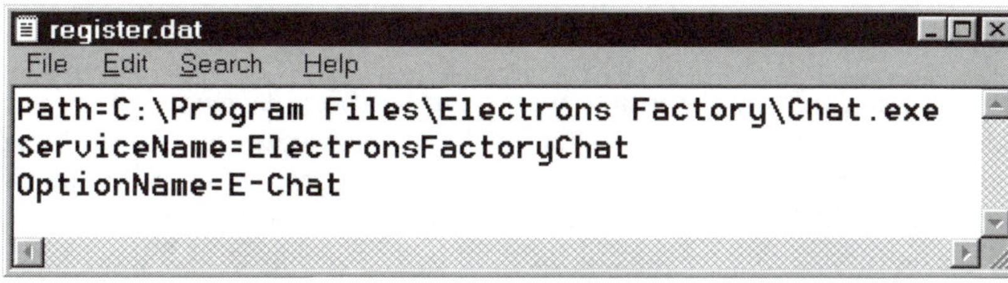

Figure 3.27 A sample register.dat file

The path information is used by Personal Communicator to locate the communication module. The service name is used to identify the type of service during service negotiation. The service option display name is shown on the service option list that is displayed when a nickname listed on the Personal Communicator is right clicked on the mouse. Users can then select the service from the displayed service option list.

If there is a conflict of the use of a similar service name by third party communication module developers during the communication module registration process, the user will be notified and asked if he wants to replace the communication module installed. To avoid such a situation, third party developers are encouraged to include their company name in the service name.

The other provision third party communication module developers need to include, is, to accept the user's ID, nickname, IP address, private session key and private session ID as the communication module's launching parameters in this respective order. In this way, the Personal Communicator can make a program call to launch the selected communication module with these parameters when the service request is successfully negotiated.

Under the online presence notification system's requirement, only communication between the Personal Communicator and the server needs to use the session key. The data packet for client-to-client communication can be sent without using encryption with the private session key. It is up to the third party communication service module provider to decide whether to use the private session key to provide higher security in their implementation. However, communication modules are encouraged to use the private session ID to identify each other instead of only using the IP address when establishing direct connection. In short, third party communication module developers are given the total freedom to implement their communication services. They do not need to use everything provided.

3.8.2.2 Example of Communication Module: Instant Messaging

A communication module for instant messaging has been developed as an implementation example. This communication module demonstrates the use of the open application interface of Personal Communicator to negotiate for communication service via the online presence notification system. It is important to note that the example is not necessarily a standard for third party developers to follow; they are free to design their own modules based on their needs.

Unlike text chat or voice chat where communication is continuous in nature, instant messaging is more sporadic. Once the recipient accepts instant messaging service from the requester, the private session ID and key for the instant messaging service are valid for the whole online session until either party ends his online session. This is to prevent the Personal Communicator from asking the recipient if he wants to accept instant messages from the sender each time an instant message is sent to him. In contrast, for real-time communication services which are continuous in nature, the private session ID and key are only valid for the particular chat session but not the whole online session. The private session ID and key are discarded as soon as the chat session is ended.

Currently, the content of each instant message is limited to 500 characters while the total size of each message packet is set at 1,000 characters. Therefore, the instant messaging module will not read anything beyond the 1,000th character. Figure 3.28 shows the instant messaging screen.

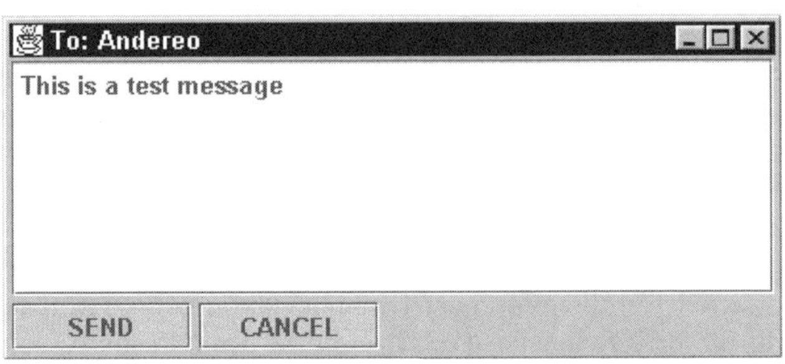

Figure 3.28 Instant messaging service

References, Links and Bibliography

[1] G. Koprowski, "Emerging uncertainty over IPv6", *Computer*, Vol. 31, No. 11, pp. 16–18, Nov. 1998.

[2] A. Durand, "Deploying IPv6", *IEEE Internet Comput.*, Vol. 5, No. 1, pp. 79–81, Jan.-Feb. 2001.

[3] T. Braun, "Internet protocols for multimedia communications, Part 1: IPng – the foundation of Internet protocols'', *IEEE Multimedia*, Vol. 4, No. (3), pp. 85–90, July-Sep. 1997.

[4] T. Braun, "Internet protocols for multimedia communications, Part 2: resource reservation, transport, and application protocols'', *IEEE Multimedia*, Vol. 4, Oct.–Dec. pp. 75–82, 1997.

[5] J. Postel, "*Transmission Control Protocol, DARPA Internet Program Protocol Specification*'', RFC793, University of Southern California, Information Sciences Institute, 1981. http://www.isi.edu/in-notes/rfc793.txt.

[6] J. Postel, "*User Datagram Protocol*'', RFC 768, University of Southern California, Information Sciences Institute, 1981. http://www.isi.edu/in-notes/rfc768.txt.

[7] J. Postel, "*Internet Protocol, Darpa Internet Program Protocol Specification*'', RFC 791, Information Sciences Institute, University of Southern California, 1981. http://www.isi.edu/in-notes/rfc791.txt.

[8] H. Schulzrinne, S. Casner, R. Frederick and V. Jacobson, "*RTP: A Transport Protocol for Real-Time Applications*'', RFC1889, Network Working Group, Audio-Video Transport Working Group, 1996. http://www.isi.edu/in-notes/rfc1889.txt.

[9] T. Berners-Lee (MIT/LCS), R. Fielding (UC Irvine) and H. Frystyk (MIT/LCS), "*Hypertext Transfer Protocol – HTTP/1.0*", RFC 1945, Network Working Group, 1996. http://www.isi.edu/in-notes/rfc1945.txt.

[10] R. Fielding (UC Irvine), J. Gettys, J. Mogul (DEC), H. Frystyk and T. Berners-Lee (MIT/LCS), "*Hypertext Transfer Protocol -- HTTP/1.1, RFC2068*'', Network Working Group, 1997. http://www.isi.edu/in-notes/rfc2068.txt.

[11] H. Schulzrinne, A. Rao and R. Lanphier, "*Real Time Streaming Protocol (RTSP)*'', RFC 2326, Columbia University, Netscape, RealNetworks, April 1998. ftp://ftp.isi.edu/in-notes/rfc2326.txt.

[12] M. Nadeau, Your Email Is Obsolete, "*BYTE Magazine*", CMP Media Inc, 1997, February 97, pp. 66–80.

[13] J.B. Postel, "*Simple Mail Transfer Protocol*'', RFC821, 1982. http://www.cis.ohio-state.edu/htbin/rfc/rfc821.html.

[14] N. Borenstein and N. Freed, "*MIME (Multipurpose Internet Mail Extensions) Part One: Mechanisms for Specifying and Describing the Format of Internet Message Bodies*'', RFC1521, 1993. http://www.cis.ohio-state.edu/htbin/rfc/rfc1521.html.

[15] K. Moore, "*MIME (Multipurpose Internet Mail Extensions) Part Two: Message Header Extensions for Non-ASCII Text*'', RFC1522, 1993. http://www.cis.ohio-state.edu/htbin/rfc/rfc1522.html.

[16] K.A. Jamsa and K. Cope, "*Internet Programming*'', Jamsa Press, 1995.

[17] M. Rose, "*Post Office Protocol-Version 3*'', RFC1225, 1991. http://www.cis.ohio-state.edu/htbin/rfc/rfc1225.html.

[18] http://www.microsoft.com/exchange, 2004.

[19] http://www.lotusnotes.com, 2004.

[20] http://www.hotmail.com, 2004.
[21] http://email.excite.com, 2004.
[22] T. Berners-Lee, R. Fielding and H. Frystyk, "*Hypertext Transfer Protocol - HTTP/1.0*", RFC1945, 1996. http://www.cis.ohio-state.edu/htbin/rfc/rfc1945.html.
[23] http://www.icq.com, 2004.
[24] http://www.licq.com, 2004.
[25] http://www.aol.com/aim, 2004.
[26] http://messenger.yahoo.com, 2004.
[27] http://messenger.msn.com, 2004.
[28] A.C.M. Fong, S.C. Hui and C.T. Lau, "Towards an open protocol for secure online presence notification", *Comput. Stand. Interfaces*, Vol. 23/4, pp. 311–324, 2001.
[29] http://www.rsasecurity.com/rsalabs/pkcs, 2004.
[30] http://www.microsoft.com/windows/netmeeting, 2004.
[31] http://www.vocaltec.com, 2004.
[32] http://www.voxware.com, 2004.
[33] http://java.sun.com/applets, 2004.
[34] ftp://gaia.cs.umass.edu/pub/hgschulz/nevot/, 2004.
[35] V. Hardman, M.A. Sasse, M. Handley and A. Watson, "Reliable Audio for Use over the Internet", *Proceedings of the INET95*, Ohahu, HI, 1995.
[36] http://www.cs.columbia.edu/~jdrosen/aisfinal/aisindex.html, 2004.
[37] J.C. Bolot and A. Vega-Garcia, "A Control Mechanism for Packet Audio in the Internet", *Proceedings of the Conference on Computer Communications (IEEE Infocom)*, San Francisco, CA, pp. 232–239, 1996.
[38] K.V. Chin, S.C. Hui and S. Foo, "Enhancing the quality of Internet voice communication for Internet telephony systems", *J. Netw. Comput. Appl.*, Vol. 21, pp. 203–218, 1998.
[39] J. Busse, D. Deffner and H. Schulzrinne, "Dynamic QoS control of multimedia applications based on RTP", *Comput. Commun.*, Vol. 19, pp. 49–58, 1996.
[40] http://www.cs.ucl.ac.uk/staff/P.Gevros/jat.html, 2004.
[41] http://java.sun.com/products/java-media/jmf/index.html, 2004.
[42] D. Wu, Y.T. Hou, W. Zhu, Y-Q Zhang and J.M. Peha, "Streaming video over the Internet: approaches and directions", *IEEE Trans. Circuits Syst. Video Technol.*, Vol. 11, No. 3, pp. 282–300, Mar. 2001.
[43] B. Girod, U. Horn and B. Belzer, "Scalable Video Coding with Multiscale Motion Compensation and Unequal Error Protection", *Proceedings of the Symposium on Multimedia, Communication and Video Coding*, NYC, NY, pp.475–482, Oct. 1995.
[44] D. Taubman and A. Zakhor, "A common framework for rate and distortion based scaling of highly scalable compressed video", *IEEE Trans. Circuits Syst. Video Technol.*, Vol. 6, pp. 329–354, Aug. 1996.

[45] S. Skenker, C. Partridge and R. Guerin, "*Specification of Guaranteed Quality of Service*'', REC 2212, Internet Engineering Taskforce, Sept. 1997.
[46] K. Nichols, V. Jacobson and L. Zhang, "*A Two-Bit Differentiated Services Architecture for the Internet*'', RFC 2638, Internet Engineering Taskforce, July 1999.
[47] L. Zhang, S. Deering, D. Estrin, S. Shenker and D. Zappala, "RSVP: a new resource ReSerVation protocol'', *IEEE Netw.*, Vol. 7, pp. 8–18, Sept. 1993.
[48] M. Furini and D.F. Towsley, "Real-time traffic transmission over the Internet'', *IEEE Trans. Multimedia*, Vol. 3, No. 1, pp. 33–40, Mar. 2001.
[49] D-N Yang, W. Liao and Y.T. Lin, "MQ: an integrated mechanism for multimedia multicasting'', *IEEE Trans. Multimedia*, Vol. 3, No. 1, pp. 82–97, Mar. 2001.
[50] D. Wu, Y.T. Hou and Y-Q Zhang, "Transporting real-time video over the Internet: challenges and approaches'', *Proc. IEEE*, Vol. 88, No. 12, pp. 1855–1875, Dec. 2000.
[51] S. Floyd and K. Fall, "Promoting the use of end-to-end congestion control in the Internet'', *IEEE/ACM Trans. Network.*, Vol. 7, pp. 458–472, Aug. 1999.
[52] J.C. Bolot and T. Turletti, "Experience with control mechanisms for packet video in the Internet'', *ACM Comput. Commun. Rev.*, Vol. 28, No. 1, pp. 4–15, Jan. 1998.
[53] A. Eleftheriadis and D. Anastassiou, "Meeting Arbitrary QoS Constraints Using Dynamic Rate Shaping of Coded Digital Video'', *Proceedings IEEE International Workshop Network and Operating System Support for Digital Audio and Video*, NYC, NY, pp. 95–106, Apr. 1995.
[54] D. Wu, Y.T. Hou, W. Zhu, H-J Lee, T. Chiang, Y-Q Zhang and H.J. Chao, "On end-to-end architecture for transporting MPEG-4 video over the Internet'', *IEEE Trans. Circuits Syst. Video Technol.*, Vol. 10, pp. 923–941, Sept. 2000.
[55] T. Turletti and C. Huitema, "Videoconferencing on the Internet'', *IEEE/ACM Trans. Networking*, Vol. 4, pp. 340–351, June 1996.
[56] Z-L Zhang, S. Nelakuditi, R. Aggarwa and R.P. Tsang, "Efficient Server Selective Frame Discard Algorithms for Stored Video Delivery Over Resource Constrained Networks'', *Proceedings of IEEE INFOCOM*, NYC, NY, pp. 472–479, Mar. 1999.
[57] N. Yeadon, F. Garcia, D. Hutchison and D. Shepherd, "Filters: QoS support mechanisms for multipeer communications'', *IEEE J. Sel. Areas Commun.*, Vol. 14, pp. 1245–1262, Sept. 1996.
[58] W. Ding and B. Liu, "Rate control of MPEG video coding and recording by rate-quantization modeling'', *IEEE Trans. Circuits Syst. Video Technol.*, Vol. 6, pp. 12–20, Feb. 1996.
[59] A. Vetro, H. Sun and Y. Wang, "MPEG-4 rate control for multiple video objects'', *IEEE. Trans. Circuits Syst. Video Technol.*, Vol. 9, pp. 186 199, Feb. 1999.

[60] T. Weigand, M. Lightstone, D. Mukherjee, T.G. Campbell and S.K. Mitra, "Rate-distortion optimized mode selection for very low bit-rate video coding and the emerging H.263 standard", *IEEE Trans. Circuits Syst. Video Technol.*, Vol. 6, pp. 182–190, Apr. 1996.

[61] J. Lee and B.W. Dickenson, "Rate-distortion optimized frame type selection for MPEG encoding", *IEEE Trans. Circuits Syst. Video Technol.*, Vol. 7, pp. 501–510, June 1997.

[62] I. Rhee, "Error Control Techniques for Interactive Low-Bit-Rate Video Transmission Over the Internet", *Proceedings of the ACM SIGCOMM*, NYC, NY, Aug. 1998.

[63] W-T Tan and A. Zakhor, "Video multicast using layered FEC and scalable compression", *IEEE Trans. Circuits Syst. Video Technol.*, Vol. 11, pp. 373–386, Mar. 2001.

[64] J.C. Bolot and V.G. Andres, "Control Mechanisms for Packet Audio in the Internet", *Proceedings of the Conference Computer Communication, IEEE INFOCOM*, San Fransisco, CA, pp. 232–239, 1996.

[65] A. Albanese, J. Blomer, J. Edmonds, M. Luby and M. Sudan, "Priority encoding transmission", *IEEE Trans. Inf. Theory*, Vol. 42, pp. 1737–1744, Nov. 1996.

[66] J.C. Bolot and T. Turletti, "Adaptive Error Control for Packet Video in the Internet", *Proceedings of the IEEE Conference on Image Processing*, NYC, NY, pp.25–28, Sept. 1996.

[67] N. Shacham and P. McKenney, "Packet Recovery in High Speed Networks Using Coding and Buffer Management", *Proceedings IEEE INFOCOM*, NYC, NY, pp. 124–131, 1990.

[68] C.W. Yap and K.N. Ngan, "Error resilient transmission of SPIHT coded images over fading channels", *IEE Proc. Vis. Image Signal Process.*, Vol. 148, pp. 59–64, Feb. 2001.

[69] I. Moccagatta, S. Soudagar, J. Liang and H. Chen, "Error-resilient coding in JPEG-2000 and MPEG-4", *IEEE J. Sel. Areas Commun.*, Vol. 18, pp. 899–914, June 2000.

[70] Y. Wang and Q-F Zhu, "Error control and concealment for video communication: a review", *Proc. IEEE*, Vol. 86, pp. 974–997, May 1998.

[71] A.H. Sadka, F. Eryurthlu and A.M. Kondoz, "Error-resilience improvement for block-transform video coders", *IEE Proc. Vis. Image Signal Process.*, Vol. 144, pp. 369–376, Dec. 1997.

[72] A.C.M. Fong and S.C. Hui, "IQCM: a robust and efficient multimedia data recovery mechanism", *IEEE Trans. Consum. Electron.*, Vol. 47, No. 4, pp. 831–837, Nov. 2001.

[73] A.C.M. Fong and S.C. Hui, "A Web-based intelligent surveillance system for detection of criminal activities", *IEE Comput. Control Eng. J.*, Vol. 12, No. 6, pp. 263–270, 2001.

[74] A.C.M. Fong and S.C. Hui, "An end-to-end solution for Internet lecture delivery", *Campus Wide Inf. Syst.*, Vol. 19, No.2, pp. 45–51, 2002.

[75] K-D Seo, S-H Lee, J-K Kim and J-S Koh, "Rate control algorithm for fast bit-rate conversion transcoding", *IEEE Trans. Consum. Electron.*, Vol. 46, No. 4, pp. 1128–1136, Nov. 2000.
[76] J.A. Sprey, "Videoconferencing as a communication tool", *IEEE Trans. Prof. Commun.*, Vol. 40, No. 1, pp. 41–47, Mar. 1997.
[77] B. Goode, "Voice over internet protocol (VoIP)", *Proc. IEEE*, Vol. 90, No. 9, pp. 1495–1517, Sept. 2002.
[78] http://rfc3261.x42.com, 2004.
[79] L.S.K. Chong, S.C. Hui and C.K. Yeo, "Towards a unified messaging environment over the Internet", *Cybernet. Syst.*, Vol. 30, No. 6, pp. 533–550, 1999.
[80] http://www.jfax.com, 2004.
[81] http://www.2bsure.com, 2004.
[82] http://www.unified-messaging.com, 2004.
[83] http://www.ipost.net, 2004.
[84] M. Hurwicz, The Universal Inbox, "*BYTE Magazine*", CMP Media Inc., September 1997. http://www.byte.com/art/9709/sec6/art2.htm
[85] S.C. Hui, A.C.M. Fong and C.T. Lau, "Unified personal mobile communication services for a Wireless Campus", *Campus Wide Inf. Syst.*, Vol. 19, No.1, pp. 27–35, 2002.

CHAPTER 4

INTERNET SECURITY

4.1 INTRODUCTION

Internet security is a critically important issue for all bona fide Internet users, whether they are private individuals, government departments, corporate entities or other kinds of organizations. This chapter describes the security issues related to the use of the Internet, as well as the current tools that are used to address those issues. The security issues of concern include the protection of data integrity when stored in a database or during transmission, the protection of systems from unauthorized access, and so on.

The Internet has grown tremendously over the past decade or so. Through its open nature, the Internet set the foundation for the global community and access to resources that millions of computer users enjoy today. However, the Internet was not designed as a secure network. Any outsider with a little know-how and some malicious intent can easily invade an Internet connection between two machines. Once inside, an intruder can get some secret information such as user account, or important data that should only be disclosed to those meant to see it.

Indeed, security issues are taken very seriously by the Internet community. For example, the World Wide Web Consortium (W3C) has provided important security considerations for the design of Web-based systems, including guidelines on Web server configuration and user access control. The International Standards Organization (ISO) ISO 7498-2 defines five important categories of security services. These are authentication, access control, confidentiality, data integrity and non-repudiation. We discuss the various security-related issues both from the users' perspectives and from the system developers' perspectives. In particular, the following topics will be presented:

- Overview of Internet security
- Practical approaches

- Authentication using biometrics — a case study with multi-view facial analysis

4.2 Internet Security — An Overview

Internet security techniques can be classified in a number of ways. From a system developer's perspective, one way is to classify these techniques into those techniques that are related to the Web server security and those that are related to software security. Software security-related techniques can be further classified into those that are related to programming and those related to communication. In general, an important assumption is made: in each organization, there are certain senior members of staff and system administrators that will have full access to all security-related components within a system. It is, therefore, generally assumed that each organization will have a (small) number of employees who can be trusted unconditionally. The reliance on the integrity of personnel cannot be underestimated. As an analogy, even the supposedly most secure prison in the world, Camp Delta at Guantanamo Bay, Cuba has had security breaches caused by a number of rogue employees. So, no matter how secure an Internet system is, it can only be as good as the level guaranteed by the integrity of staff that have privileged access to the system. This underlines the fact that any system can only be as good as its weakest component.

From a user's perspective, typical security requirements include the protection of the integrity of valuable data, as well as the availability of data. The availability of services and data has recently become an important issue following a spate of service denial attacks. Another important issue is access control. The purpose of this is to ensure that not only authorized persons can gain access, but also that the level of access that each individual is granted must depend on the person's status within the organization. A closely related issue is user authentication. Having set up the different levels of access for different individuals, how do we ensure that a person is who he/she claims to be? Is a person claiming to be X (and wantings to gain access to authorized data authorized) really the person X stored on the records? User authentication is used to answer these questions. Security issues and requirements viewed from the two different perspectives are not mutually exclusive. Indeed, they are highly related.

4.2.1 Web Server Related Security

To protect the integrity of system services and data, the Web server itself must be secure. This security encompasses both physical security and a carefully managed and customized access control system for different users of different categories in terms of their privilege status. Such a secure Web server is only a necessary condition for maintaining the validity of all cryptographic measures described later in this section. However, it is not sufficient on its own.

The first necessary condition that must be satisfied is that the Web server must be physically safe. It seems to be obvious that the server must be made inaccessible by unauthorized individuals. However, this is sometimes overlooked in practical deployments. The Web server must be stored in a locked enclosure. There must also be a proper procedure for key management.

The second necessary condition concerns Web server configuration. The configuration of access rights to critical files stored on the Web server is extremely important. Only operating systems with at least a built-in password-based access control facility can be employed. Of course, more security features are desirable because password-based systems can also be compromised, especially when authorized users are not careful with the protection of their passwords.

Also, a procedure must be in place to ensure that an ordinary local user cannot, whether intentionally or otherwise, change the Web server configuration file or the document tree in such a way that a security loophole could be created. Hence file permissions in the server's directories must be carefully maintained in order to ensure that unauthorized local users cannot indirectly gain access privileges to install application software and access critical information (such as the encryption key). For instance, one may create a "www" group for which only trusted Web authors are added. The document root must then be made writable only by members of this group. To increase security further, the server root where vital configuration files are kept should be made writable only by the designated Web administrator.

Another important aspect is access control. User accesses to the Web server, either locally via the Intranet or remotely via the Internet, must be controlled carefully. Although eavesdropping by internal staff is possible, it is not a major concern in this study. What we want to say is that for local users, we could have a common (i.e. same for all local users) and standalone (i.e. separate from policies for other users) security policy that applies locally on the LAN. Control over remote users can be achieved through configuration set-ups of the Web server, use of firewalls or proxy servers.

Access control capability provided by Web servers to authorize access from other hosts as specified in terms of Internet protocol (IP) addresses and/or host names can be employed in this regard. For example, Figure 4.1 shows the Netscape Enterprise Server's interface for access control definition. IP address restriction can be made much safer by running the Web server behind a firewall machine that is capable of detecting and rejecting attempts at spoofing IP addresses [1]. The firewall machine intercepts packets originated from the outside world that attempt to disguise as messages from trusted machines on the internal network.

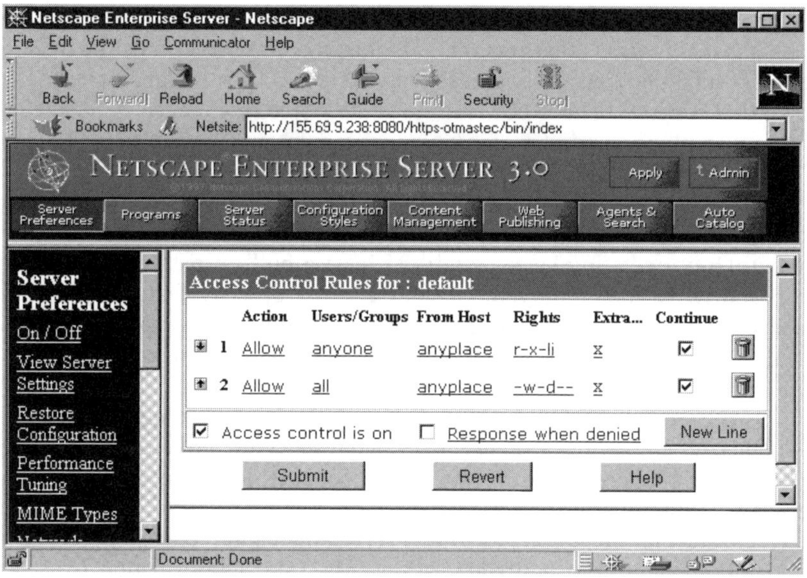

Figure 4.1 Netscape Enterprise Server 3.0's interface for access control definition

Extra care must be exercised in cases where a browser is fetching documents through a proxy server. In this instance, the Web server will only recognize the IP address of the proxy in such a set-up, but unable to identify the real user. This means that if the proxy is in a trusted domain, anyone can use that proxy to access the Web site. Unless a particular proxy can be trusted to do its own restriction, it becomes imperative not to add the IP address of a proxy (or a domain containing a proxy server) to the list of authorized addresses.

Similarly, it is necessary for the system administration to be aware of "DNS (Domain Name Server) spoofing" when using a restriction by host/domain name. In such an attack, the Web server is temporarily fooled into thinking that a trusted host name belongs to an alien IP address. To lessen the risk, the server should be configured to do an extra DNS lookup for each client. After translating the IP address of the incoming request to a host name, the server uses the DNS to translate from the host name back to the IP address. Permission to access is granted only if the two addresses are identical.

One of the simplest and most widely used methods of access control is the use of passwords. Good password policies and control should be implemented. Users with login privileges should be made to choose good passwords. A poorly chosen password, such as one that can be guessed easily from the user's background, is likely to be used by unauthorized users. The critically important administrator's password must be carefully controlled. A policy for regular changes to password must be enforced. In summary, the following guidelines are recommended for defining passwords.

- Passwords should be at least seven alphanumeric characters in length, including different types of characters (a mix of digits, upper and lower case letters, and special characters such as "#", "$", "%"), which means they should preferably be case sensitive.
- Proper names, common names, usernames, the names of pets and other information that could be easily deduced (birthdays, and so on) should be avoided. In fact, it is preferable not to use any words that can be found in a dictionary. Passwords that are phonetically difficult to pronounce are even better.
- A regular password change should be enforced (such as once in every 3 months). An even better practice is to change the password in a "sporadic" fashion within these regular intervals.

These guidelines are intended to hinder hackers who employ password-guessing programs such as AntiCrack [2] to break in by means of brute force. Of course, password-based access control systems have some fundamental flaws. For example, passwords can be forgotten and are not foolproof. These problems can be addressed by the use of biometrics discussed later in this chapter.

Another security aspect, which is easily overlooked, is that unused services should be turned off whenever possible. For example, if there is no need to run a File Transfer Protocol (FTP) service on the Web server host, the FTP software should either be removed or disabled. This should also be the case for other Internet services such as Trivial File Transfer Protocol (TFTP), Gopher, Network File System (NFS) and anything else that may be present but not necessary for the server's proper operation. Regular checks on servers that may be lurking should be carried out.

In addition, both system and Web logs should be checked regularly for suspicious activity. Normal checks include the identification of accesses involving system commands or extremely long lines in Universal Resource Locator (URL) requests. The former may indicate an attempt to trick a Common Gateway Interface Common Gateway Interface (CGI) script into invoking a system command; the latter may be an attempt to overrun a program's input buffer. Particular attention should be focused on repeated attempts to access password protected documents. These could be symptomatic of someone trying to guess a password. If this really happens, the system should deny further accesses from these IP addresses or domain names by re-configuring the Web server, the firewall machine or the proxy server.

Finally, another security aspect that is easily overlooked is to evaluate the latest service packs before installation and widespread use. Newest update packs of system software that are regularly distributed by vendors should be carefully analysed with respect to security implications before use. These updates may be used to introduce new features, enhance performance or rectify existing bugs. However, these improvements may be achieved at the expense (albeit unintentionally and indirectly) of security needs.

4.2.2 Software Security

Authentication, confidentiality and data integrity can be addressed by well-studied cryptographic techniques [3, 4], which are described in more details later in this chapter. When using such techniques, it is recognized that information en route on the Internet can pass through numerous computers before it reaches its destination. A malicious user of any of the intermediary computers can monitor the Internet traffic en routes, eavesdrop, intercept, modify or replace the data along the way. Cryptographic techniques can be used to protect this data. There are two kinds of key-based encryption systems: symmetric and asymmetric systems [3].

Symmetric encryption uses the same secret key to encrypt and decrypt a message. Two users wishing to communicate in confidence must agree in advance, via a *secure* channel, to maintain a common secret key and each entity must trust the other in not divulging the key to a third party. In addition to the key distribution and management problem [4], the fact that multiple parties know the key implies that authentication by means of knowledge of the key becomes very difficult [2]. Non-repudiation issues are almost impossible to address when any one in the group could edit or replace messages in transit from the source to the intended recipient.

Asymmetric encryption separates the functions of encryption and decryption by having a complementary pair of keys; usually referred to as the private key and the public key. The private key is normally kept secret while the public key does not need to be kept secret. This two-key approach simplifies key management by minimizing the number of keys to be stored and allows the distribution of public keys via unprotected open networks. The difficulties in meeting authentication and non-repudiation needs in symmetric systems as described earlier vanish in the asymmetric system. However, asymmetric systems are generally much less efficient in terms of computation.

In an asymmetric encryption system, if Alice wishes to securely transmit messages to Bob, Alice must first obtain the certified public key from Bob. This is carried out in the open. Alice will then encrypt the message using Bob's public key. Since the complimentary private key is necessary for decrypting the message, it can only be decrypted and read by Bob, the only one in possession of the private key required. Even the originator of the message cannot decrypt it once encryption has taken place.

Different asymmetric cryptographic algorithms based upon different mathematical problems exist. RSA is an example of *the Integer Factorization problem* [5, 6]. Examples of *Discrete Logarithm problem* [3] include the ElGamal's system and its variations such as the one employed in the Digital Signature Algorithm (DSA). A DSA equivalent, Elliptic Curve Digital Signature Algorithm (ECDSA), has been proposed based on the *Elliptic Curve Discrete Logarithm problem* [5]. Studies on the advantages of the various cryptosystems have been reported by Schneier [3]. For example, RSA is easy to implement, ElGamal works well for encryption, and DSA is great for digital signatures.

An analysis of the security requirements reveals that a simple cryptographic protocol turns out to be sufficient for most applications. The protocol is carefully designed by following the design guidelines suggested in [7].

4.3 PRACTICAL APPROACHES

This section describes popular and practical approaches for ensuring a high level of security. Practical approaches revolve around security issues related to access security (Section 4.1) and transfer security (Section 4.2). The important topic of cryptography is described in more details in Section 4.3. Section 4.4 highlights some commercial products.

4.3.1 Access Security

Access security concerns user authentication and access to the resources available to authorized users. Popular methods of verifying the identity of users who attempt to access restricted resources include the use of cookies, digital certificates and username/password authentication.

Cookies [8, 9] are pieces of information sent from the server to the browser, which the browser sends back with every request from the Web user. For obvious privacy and interaction reasons, cookies are only sent back to the server that created them. Cookies are often used to store an ID number that the server uses to reference a database record containing information about a particular user. Cookies have the advantage of being changeable by the host server, so that they can carry a variable amount of information. Another advantage of cookies is that they are user-transparent. However, since cookies can also be changed easily by the receiver or used by any other person who can intercept the cookie during transmission, they do not provide a high degree of security. Cookies are, therefore, used mainly for lower-security applications.

Digital certificates [8, 10] are issued by certificate authorities (CAs), signed with private keys, and verified with public keys. Like cookies, digital certificates are character strings delivered to the user's computer. Unlike cookies, information about the sender contained in digital certificates is encrypted and can only be decrypted by a trusted CA. Digital certificates can be used as proof that the data is in fact coming from the browser to which a unique certificate was assigned. Digital certificates verify that the computer, browser or user to which they are attached is the original item that they were issued to, not an illegal copy. Therefore, digital certificates have the advantage of being highly secure. Like cookies, digital certificates are also highly user-transparent. However, they do not necessarily identify the person sending a request. They may only identify the sending computer or software program. In addition, one disadvantage of digital certificates is that the procedure for the sender to get a certificate and the receiver to set up a verification process can be long and costly.

As mentioned above, username and password authentication is one of the most widely used access control methods. In this form of security,, users are required to provide a pre-authorized name and password before they are given access to resources on the host computer, or perform actions that can only be performed only by those authorized users. This method has the advantage of being portable, for example, resources can be accessed from anywhere on the Internet. However, it cannot offer true security. Because the username and password are transmitted across the Internet in the clear, any eavesdropper can obtain them without much effort. Encrypting the username and password before transmitting them can increase the security of this method.

4.3.2 Transfer Security

Giving authorized users access to information is only the first part of the information distribution process. The information such as control information or media stream must then be capable of transfer across the Internet back to their computers without any unauthorized person seeing what is in the stream. *Transfer security* relates to the protection of information transferred across the Internet. It involves various encryption techniques such as *public-key and private-key cryptography*, *Secure Sockets Layer* (*SSL*) *encryption*, and *digitally signed encryption*.

Cryptography [11, 12] is a practical application using reasonably difficult mathematical concepts. The idea is to transform some data into a form that is as close as impossible to understand without the appropriate knowledge (a key). It aims to ensure security by keeping information concealed from anyone for whom it is not intended, even those who can access the encrypted data. *Public key* and *private key* [11] are two broad categories of encryption algorithms.

Private-key encryption, or symmetric encryption, depends on a secret key. The same key is used for encrypting and decrypting data. Private-key encryption is a useful tool when you need to transfer secure data over an insecure network. The advantage of secret-key encryption is that it is generally much faster than public-key encryption. The main problem of secret-key encryption is getting the sender and receiver to agree on the secret key without anyone else finding out. This needs to find a method by which the two parties can communicate with each other without fear of eavesdropping.

Public-key encryption, also called *asymmetric encryption*, uses two closely related keys together. These two keys — the public key and the private key — are often called a *key pair*. Typically, the public key is used to perform the encryption, and the private key is used to perform the decryption. While the public key can be published, the private key must be kept secret. Public-key encryption is much slower than private-key encryption. It can take 1,100 times as long as private-key encryption to encrypt or decrypt data. To provide an equal level of security, public-key encryption also requires keys up to 10 times as long as those for private-key encryption [10]. Since public-key encryption is so slow, in situations where two different parties want to exchange data using the vastly more efficient private-key encryption, the public-key encryption is used first to pass a secure one-time-use private key to both parties.

SSL [13, 14] is the Internet security protocol for point-to-point connection. SSL provides data encryption, server authentication, message integrity and optional client authentication for a TCP/IP connection. It is a popular encryption technology used to establish a secure Web connection. Using SSL, the user and host computer must have an initial negotiation to establish an agreement on the method of encryption. Subsequently, all information is transferred across the Internet after it has been encrypted with the agreed-upon method. Even though an intruder might be able to intercept the data, they would not be able to decode it. The main problem with SSL is that although some information can be sent through it using Transmission Control Protocol (TCP), streamed video cannot. Streamed video is sent over the user datagram protocol (UDP). UDP is outside the protection of SSL.

Today, the growth and increasing complexity of Internet business users are demanding a shift from defensive to enabling technologies, and from single-dimensional to multi-dimensional security techniques [15]. Multi-dimensional security uses different encryption methods and mechanisms to create as comprehensive a security system as possible. For example, digital signature and data encryption can be combined to provide multi-dimensional security for data. It refers to the process of dual-encryption — sealing data already encrypted with another encryption method. Thus, the computer receiving the information can verify the source of the information. However, the problem with using digital signature and encryption with real-time video streams is that a video stream is a continuous flow of smaller packets. The additional load added to each packet by the digital signature will slow down the transfer and processing significantly. The computer receiving such a secured video stream is necessarily powerful and connecting to a dedicated high-bandwidth Internet to be able to play out the video stream without significant degradation in quality.

4.3.3 Cryptography

Cryptography [3] refers to the art and science of keeping data content secure. The original text, known as plaintext, is converted into a coded equivalent called *ciphertext* via an encryption algorithm. The ciphertext is decoded using decryption algorithm to turn it back into plaintext. As mentioned above, there are two cryptographic methods: symmetric key [16] and public key [17].

In conventional cryptography, also called *secret-key or symmetric-key encryption*, both the sender and receiver use the same key to encrypt and decrypt data. Figure 4.2 illustrates the symmetric-key method. Compared to the public-key approach, the symmetric-key technique is faster, but it has the persistent problem of key distribution's security. For a sender and recipient to communicate securely using conventional encryption, they must agree upon a key and keep it secret between themselves. If they are in different physical locations, they must trust a courier or some other secure communication medium to prevent the disclosure of the secret key during transmission. Anyone who intercepts the key in transit can later read, modify and forge all information encrypted or authenticated with that key. In many situations, it is often hard to transmit the secret key securely to the recipient. Nevertheless, symmetric-key algorithm such as the Data Encryption Standard (DES) [18] is a popular data encryption algorithm.

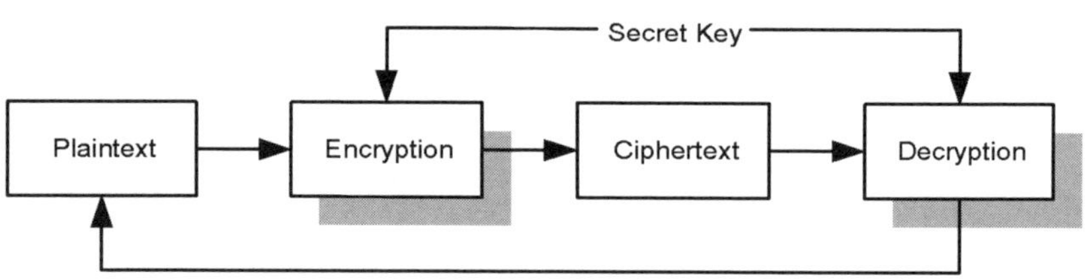

Figure 4.2 Symmetric-key encryption

The problems of key distribution are solved by public-key cryptography. Public-key cryptography as shown in Figure 4.3 is an asymmetric scheme that uses a pair of keys for encryption: a public key, which encrypts data, and a corresponding private, or secret key for decryption. Some algorithms even support the reverse usage of public key and private key for encryption and decryption. For a public-key encryption scheme to work, it must be computationally infeasible to deduce the private key from the public key. Each recipient has a private key that is kept secret and a public key that is published for everyone. The sender looks up the recipient's public key and uses it to encrypt the message. The recipient uses the private key to decrypt the message. The benefit of public-key cryptography is that it allows people who have no pre-existing security arrangement to exchange messages securely. The need for the sender and receiver to share secret keys via some secure channel is eliminated; all communications involve only public keys, and no private key is ever transmitted or shared. The RSA (named after its inventors, Ron Rivest, Adi Shamir and Leonard Adleman) [19] is an example of a popular public-key cryptography solution.

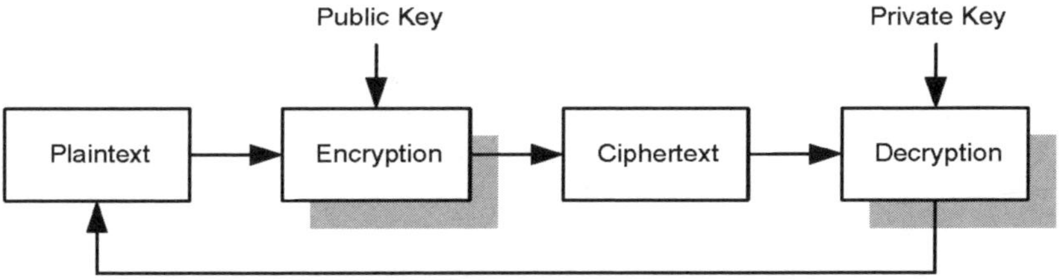

Figure 4.3 Public-key encryption

While the public-key cryptography method solves the problem of secure key distribution, it is relatively more computationally intensive. As a result, the hybrid encryption scheme, which combines the use of both the symmetric-key and the public-key techniques can be created. The purpose is to combine the convenience of public-key encryption with the speed of conventional encryption. Conventional encryption is about 1,000 times faster than public-key encryption. Public-key encryption in turn provides a solution to key distribution and data transmission issues. Used together, performance and key distribution are improved without any sacrifice in security. The Pretty Good Privacy (PGP) [20] is a popular encryption program that uses the hybrid encryption scheme.

4.3.3.1 Practical Security Mechanisms

Security mechanisms are often incorporated to provide endpoint-to-endpoint security at protocol level. These mechanisms help to prevent spoofing, service requests from unauthorized users and other illegal operations. The mechanisms for establishing communication between two parties are as follows:

- *Authentication* — hybrid cryptographic scheme is used during the authentication process. Passwords are encrypted before they are delivered over the network. This prevents password sniffing, which hinders the impersonation of users. Authenticated users are then issued with a session key, a session ID and an increment constant for the session ID by the server via the public-key encryption scheme.

- *Packet encryption* — each authenticated user is issued a session key to communicate securely with the server using the symmetric encryption scheme. Only the packet data will be encrypted. The packet header, which consists of a packet marker, a version number and the user's ID, will not be encrypted. Each session key is valid only for the online presence duration. A time-out mechanism is incorporated to cater for the situation of abnormal disconnection of users. For client-to-client communication, the private session key is used for the encryption. Each private session key is only valid for the particular communication session.
- *Session identification* — each authenticated user is issued with a numeric session ID along with a random increment constant for the session ID securely by the server during authentication. Each pair of session ID and increment constant is privately shared between each client and the server. The increment constant is used to increase the value of the session ID for each connection established between the client and the server. Communication between the client and the server will be recognized based on the proper increment of the session ID. Since packet data are encrypted with the session key, the value of the session ID is secured from viewing and manipulation by a third party. The presence of the session ID helps prevent a replay attack. If any third party manages to intercept the encrypted data at network level and replays it, the presence of an older session ID will enable the client or the server to detect the replay attack. For client-to-client communication, the private session ID is used during connections establishment. Each private session ID is only valid for the particular communication session.
- *Service negotiation* — users need to initiate real-time communication service with other users via the server. This eliminates fake requests, as the client application will only listen for communication requests via the server. If the recipient accepts the request, it will issue a private session key and session ID for the particular request to the requester via the server. The requester can then use the private session key and session ID to communicate with the recipient directly.

Public-key encryption is good for secured key distribution over the network. However, it is many times more computationally intensive than the symmetric encryption scheme. Therefore, it is not suitable for the use of real-time communications data encryption. The hybrid cryptographic scheme described above uses public-key encryption to distribute session keys and uses the distributed session keys to encrypt data using symmetric encryption. Therefore, the hybrid encryption scheme offers improved performance for real-time data transfer without compromising security.

Figure 4.4 shows how session keys are transferred over the network by the public-key encryption scheme. Session keys are not transmitted over the network once they are securely distributed by public-key encryption scheme. In addition, a session key is valid only for the service session and unique between user pairs. The session keys are then used for encrypting service requests and messages. Figure 4.5 shows the symmetric encryption scheme using session key.

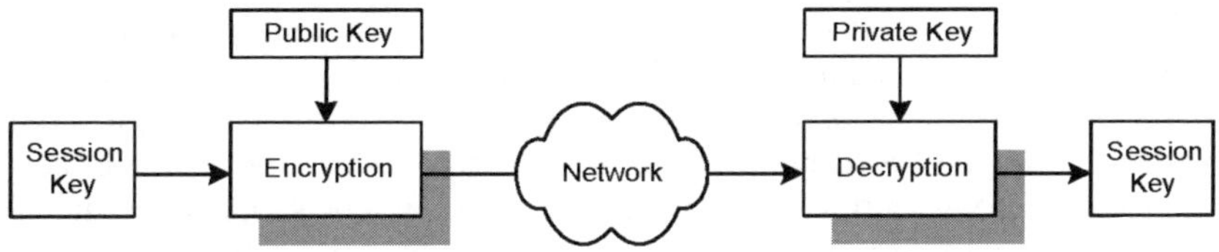

Figure 4.4 Session key distribution using public-key encryption scheme

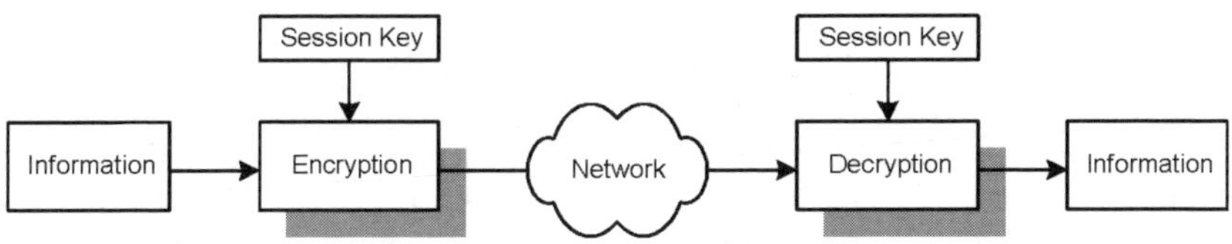

Figure 4.5 Data encryption with symmetric encryption scheme using session key

However, these cryptographic-based security mechanisms are not server friendly. They can even slow down large systems and servers that need to serve many concurrent users. As a result, many non–security-critical systems either use nothing or simple encryption algorithms to encrypt a user's password for transmission over the network. This makes such systems very insecure and they are vulnerable to attacks. One approach around this problem is to use a public-key encryption scheme for password encryption and session key distribution. Since the public-key encryption is relatively computationally intensive, symmetric encryption is also used with public-key encryption to form a hybrid cryptographic authentication protocol to provide good security during authentication. In addition, a relatively secure system should offer data encryption using a session key. It can also use a special session ID scheme for communication between client and server and between client and client. This prevents client programs from accepting data packets from unauthorized users or strangers. This helps to prevent message spoofing, anonymous messaging and fake communication requests or request replay in the system.

4.3.4 Commercial Solutions

Many vendors have addressed the security issues pertaining to Web applications. Examples include *Netscape Security Solutions* [21], *Microsoft Security Advisor Program* [22] and *Web Based Documentation - Security Issues* [23]. Besides these solutions for normal Web site security, there are also other commercial products that provide encryption services for data transferred over the Internet. SSL and Secure Hypertext Transfer Protocol (SHTTP) are two popular schemes.

SSL is a low-level encryption scheme used to encrypt transactions in higher-level protocols such as Hypertext Transfer Protocol (HTTP) and FTP. The SSL protocol includes provisions for server authentication, encryption of data in transit, and optional client authentication. SSL is currently available on several different common browsers. It is also available on server software from major vendors. However, the usage of SSL requires every legitimate user of the customer support system to register with a third party vendor at a cost, and be subjected to the 40-bit restriction imposed by the US government. This additional cost is unnecessary as all legitimate users of the customer support system are known to the system administrator of the server.

SHTTP is a higher-level protocol that only works with the HTTP protocol, though it is potentially more extensible than SSL. Currently, SHTTP is implemented for the Open Marketplace Server on the server [24], and SHTTP Mosaic [25] on the client side. The main drawback of using these off-the-shelf security products for Web applications is that they require the correct combinations of compatible browsers and servers to operate.

4.3.4.1 Application Scenario

In this scenario, we employ asymmetric encryptionbased techniques to satisfy the authentication and non-repudiatory security needs, and use symmetric encryption to meet confidentiality needs. An off-the-shelf public domain software package [26], which provides RSA asymmetric-key encryption, is employed in order to reduce development time. We consider RSA to be a well scrutinized algorithm with no known efficient general method of breaking [27], and that a 512 bit key is sufficiently secure for most applications [26, 27]. The software package used provides the functionality for message encryption, digital signatures and data compression. It also handles exceptions and avoids the usage of parameters that may result in a weak system.

Figure 4.6 shows a simple cryptographic protocol design. The protocol focuses mainly on user authentication and server authentication. Figure 4.7 shows the software architecture for implementing the protocol. In this system, besides the Web server security considerations discussed previously, *ActiveX* and *CGI* programs are used to embed cryptographic operations in the application layer. Ideally, these cryptographic functions should be developed in-house in order to have full control and assurance of the source code, although the use of properly licensed source-code packages may be functionally acceptable.

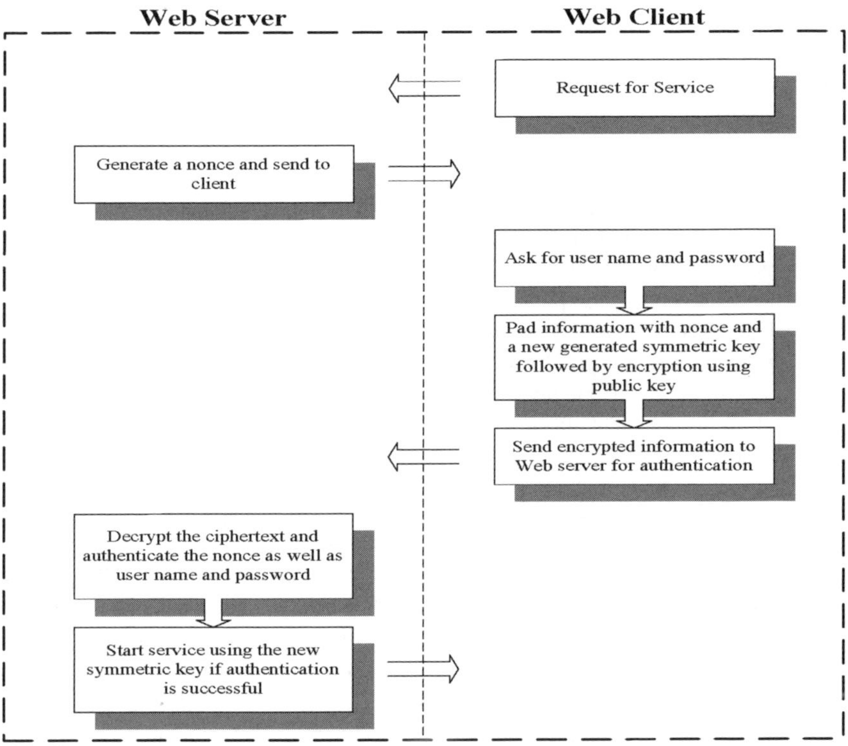

Figure 4.6 A user authentication protocol

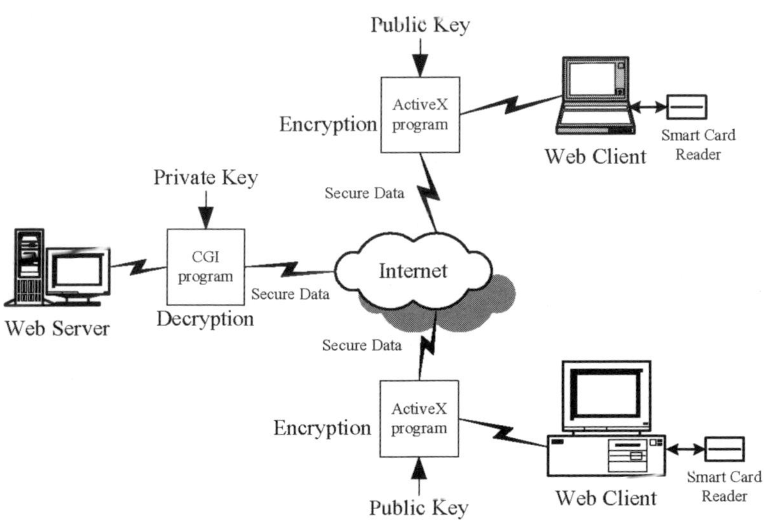

Figure 4.7 Software structure for implementing the cryptographic protocol

The process of normal client access is outlined as follows:

- When a user wishes to access the server via a Web browser, the server will first send a plaintext 64-bit random number n_s (as the nonce[1]) to the Web client and start the *ActiveX* program on the client side. The *ActiveX* program should be pre-installed on the Web client side; otherwise some alternative form of server authentication by the client must be put into place if downloading is used instead. This can be achieved, for example, by accompanying the program with a digital signature produced using the server's private key [3].
- The *ActiveX* program on the Web client side will require a name and password to be supplied by the user. In the current prototype, these are then combined (i.e. Exclusive-ORed or padded) with n_s before being encrypted using the server's public key and transmitted to the Web server for authentication. If deemed necessary, the client can also generate a random session key k_s for symmetric encryption of subsequent data (i.e. DES in our application) and include it with the encrypted reply to the Web server. For an even more robust solution, we recommend the use of a smart card to store the password and the server public key. This smart card must be physically present during each access to the server.
- The *CGI* program in the Web server will decrypt and authenticate the user before allowing access to the various server functions. If the nonce, user's name and password are all correct for the session, the user will be allowed to use the system continuously until the session is terminated by either the user or the system.
- If necessary, all subsequent data transmitted across the network between the client station and the Web server in the session can be protected by padding with n_s before symmetrically encrypted using the key k_s. This ensures the confidentiality and integrity of the data en route on the Internet.

The nonce is used mainly to confirm "freshness" in the current communication session in order to prevent replay attacks. This nonce will be discarded and the session terminated. The Web server will deny any access request without a correct nonce. In this protocol, actual user authentication is carried out by means of password authentication. Confidentiality and integrity of en route data are achieved using n_s in combination with k_s as a symmetric encryption key, both of which vary from one session to another. The ability to retrieve k_s provides the server the confirmation of data origin authentication. In the event of interruption in communication, accesses by the same user's IP address or domain name can be detected and denied. A new session must therefore be initiated. This, along with the use of the nonce, the server public key and the correct password will prevent

[1] A "nonce" is an entity to ensure freshness, usually used for the prevention of attacks by replay and message linkage.

the user from repudiating actual usage. Charges to usage of the system can then be achieved by employing system tools available on the Windows NT server software installed on the customer support server.

This implementation chooses to compile both the *ActiveX* and *CGI* programs (written in Visual C++) into binary code before deployment into the system. Hence, all encryption operations are embedded in the program, although the cryptographic algorithm can still be changed when necessary. It must be emphasized that since the server's public key is being embedded in the binary code of the *ActiveX* program, it is possible for a malicious attacker with access to the client machine to replace the program on the client machine with one that contains his own public key. He can then trick the client into releasing the password by impersonating the server. This model therefore assumes that all relevant staff members who have access to the client machine are trustworthy in this regard and that the machine is maintained by the same group of people. Otherwise, the server's public key must be placed on some removable media or device (such as a smart card). The current model permits changes to the public- and private-key pair on the Web server side by a recompilation of the client program with the new Web server's public key. The new executable program, accompanied by a digital signature produced using the original private key, can then be conveyed to the client over the Internet.

4.3.4.2 Other Practical Issues with Cryptographic-based Approaches

A strong cryptographic algorithm is powerful when it is used correctly, but it is not a panacea [4]. Focusing on the cryptographic algorithms while ignoring other aspects of security is like defending a house with strong walls, strong doors, strong window frames but without protection over the chimney. Attackers can often find simple loopholes to get around the algorithms if the system is not designed carefully. The following are further pitfalls that may be exploited by smart attackers.

While encryption is very difficult to attack, CGI scripts are a major source of security voids. CGI scripts must be written with just as much care as the server itself. Normally, compiled languages such as C/C++ can give further deterrents to potential attackers except for the most determined. Much more effort and technical expertise would be required to perform a reverse engineering process to identify the functionality of the software. However, it is also much more laborious to detect implementation defects that can induce a security void.

On the other hand, CGI scripts can also be written using scripting languages (e.g. C shell script) instead of compiled languages. In this case, it is very difficult to write a CGI script of arbitrary complexity that completely avoids dangerous constructions. For example, it is extremely easy to send data to system commands and capture their output in a scripting language, yet the invocation of system commands from within a script is a major potential source of security problems. For this reason, using a scripting language for writing CGI scripts is not recommended unless it performs a very trivial and straightforward function.

In the system described above, all the C/C++ programs are written to perform only designated functions in the simplest and most direct way. These programs are compiled and stored in the *cgi-bin* or *wincgi-bin* directories with carefully tailored access control. In addition, they are designed and implemented carefully to avoid a set of common unsafe mistakes. In particular, these programs must:

- Make assumptions about the size of user inputs and allocate fixed memory for these inputs. When there is a need to receive the user's inputs (such as the password in some CGI programs), the program would use previously fixed and defined memory for storing these inputs so that a memory overflow exception will be activated if these limits are breached.
- Prevent passing unchecked remote user inputs to a shell command. All the CGI programs in the system do not use or start any shell commands by themselves or through user inputs. This prevents the potential serious consequences that can arise if commands are allowed to execute freely.
- Ensure erasure of logs, passwords and other critical information from the memory after usage to avoid unintended covert channels. All CGI programs will reset dynamically requested memory blocks to a known form before releasing it so that no useful information is left in memory.
- Avoid giving any explicit clue that may lead to reverse engineering. This system combines many separate CGI programs into one and controls the program by using different parameters. This minimizes the possibility that an individual CGI program is giving sufficient explicit clues for tracing.
- Avoid giving explicit warning messages in a program for detecting the password. This is mainly to hinder attackers who intend to employ a bypass-password-testing attack to circumvent the password authentication process.

4.4 SECURITY FOR JAVA — AN INTERNET JAVA PHONE EXAMPLE

The Java programming language is becoming increasingly popular for a variety of reasons, such as portability due to its platform-independent nature. However, there are important security issues that pertain to the use of Java. The purpose of this section is to outline the security issues and to explore ways around the problems without jeopardizing security. In particular, the purpose is to identify ways to safely overcome the restrictions imposed by the use of Java.

Java programs can be classified into two distinct types — applications and applets. Java applets are particularly interesting. These are programs that can run from inside a Java-compatible web browser. Applets allow class files to be automatically downloaded from a Web page and run on the client machine. This allows users to execute code over the Internet without having to pre-install the program on their machine. They also offer developers the flexibility to distribute and upgrade their existing code over the Web.

However, this flexibility comes at a price. As applets are designed to be downloadable and executable across a network, certain security restrictions are placed to restrict what they can do on the client system. These restrictions only apply to applets but not to applications. Some of these restrictions include the opening of new network connections and accessing local files. To enforce these security restrictions, Java has a comprehensive security architecture that includes a customizable "sandbox" in which Java applets must run in. This protects users from downloading any malicious programs that may be hidden. The applet model is much safer compared to ActiveX [28] because an ActiveX control has no limitations on its behaviour once it is invoked [29]. Java users, on the other hand, are able to run untrusted applet code quite safely.

However, some applications, such as an IJPhone [30, 31] applet must run outside the limitations of the sandbox in order to function properly. Hence, ways must be found to overcome these limitations and not compromise the security of the client system. Two of the major applet restrictions are file access restrictions and network restrictions. File access restrictions do not allow applets to read or write to local files that exist on the client's terminal. This is a potential problem for the IJPhone because native library files (e.g. DLL files) must be loaded before audio recording and playback can be performed under Java. In the case of an application, these files are stored locally and are loaded directly into memory. The applet must therefore find a way to read and load these native library files, given the file access restrictions.

Network restrictions dictate that applets can only make network connections back to the Web server they were loaded from. An applet may not listen for incoming socket connections, nor can it listen for datagrams from anywhere but its home server. It also can only send datagrams back to its home server. This issue must be overcome because the audio communication must take place between two client applets directly, without any server intervention in order to avoid excessive packet delay.

The security restrictions imposed on Java applets vary from browser to browser. Netscape Navigator, for example, has a very tight security model, although Sun Microsystem's HotJava browser allows some of the security restrictions to be switched off. Microsoft Internet Explorer (IE) supports several security models, from completely relaxed (no restrictions) to completely secure (will not download and run applets at all). The security restrictions described above must be overcome in order to give the end-user both the functionality of an Internet telephony application, along with the convenience and ease-of-use of Java applets.

4.4.1 Java Security Architecture

The Java language was designed for networked environments, and security becomes an important issue as potentially harmful programs can be downloaded and executed on a computer connected to the network. Therefore, Java has a comprehensive security architecture that protects users from running hostile programs from untrusted network sources.

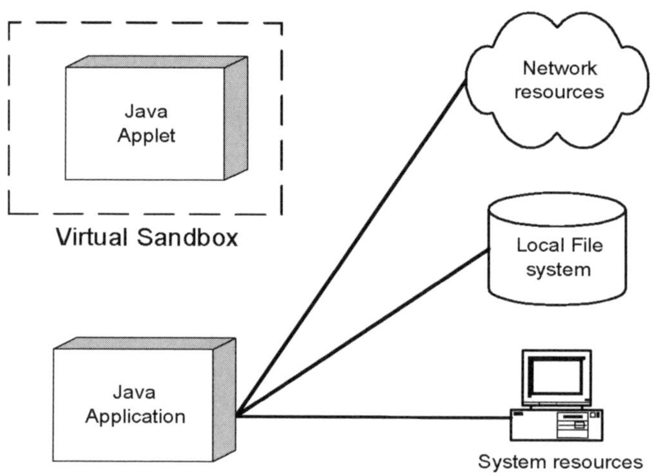

Figure 4.8 Java sandbox security model

The security architecture makes use of a "sandbox" security model where Java applets can only operate within. Java applets can perform any task within the sandbox, but cannot do anything outside of it. In comparison, Java applications operate entirely outside the sandbox and have full access to network resources, the local file system and other system resources. Figure 4.8 shows the sandbox security model. Traditionally, users enforce security on their own systems by only installing software from trusted software companies or developers. However, once the software has been installed, it can have full access to system resources like memory and file systems. Hence, it could potentially introduce viruses, Trojan horses [32] and other harmful programs to damage the system. Using the sandbox security model allows users to download programs from any source, allowing the sandbox to restrict what tasks the program can perform. There is no need to determine whether the code can be trusted or to scan for viruses [33].

The sandbox security model makes use of three mechanisms to enforce security — the Class Loader, the Bytecode Verifier and the Security Manager. Used together, these security mechanisms perform load-time and run-time checking to restrict file system, network and web browser access. All the security checks are executed on the client system. Each of the mechanisms depends on some part of the others and each part must perform its function properly for security to be enforced properly.

Figure 4.9 shows an example of how the security model is applied in the case of a Java applet downloaded across the Internet from a Web page on a remote web server. After the applet is downloaded, it is loaded by the Class Loader along with the core Java class files residing on the client's web browser. When the applet is successfully loaded, the class files of the applet go through the Bytecode Verifier to check the integrity of the applet's bytecodes. The verified bytecodes are then passed through the Security Manager to check if the applet tries to execute any restricted operations. Once the applet passes all these tests, it is then allowed to run on the client system through the local execution engine (usually the Java Virtual Machine of the Web browser).

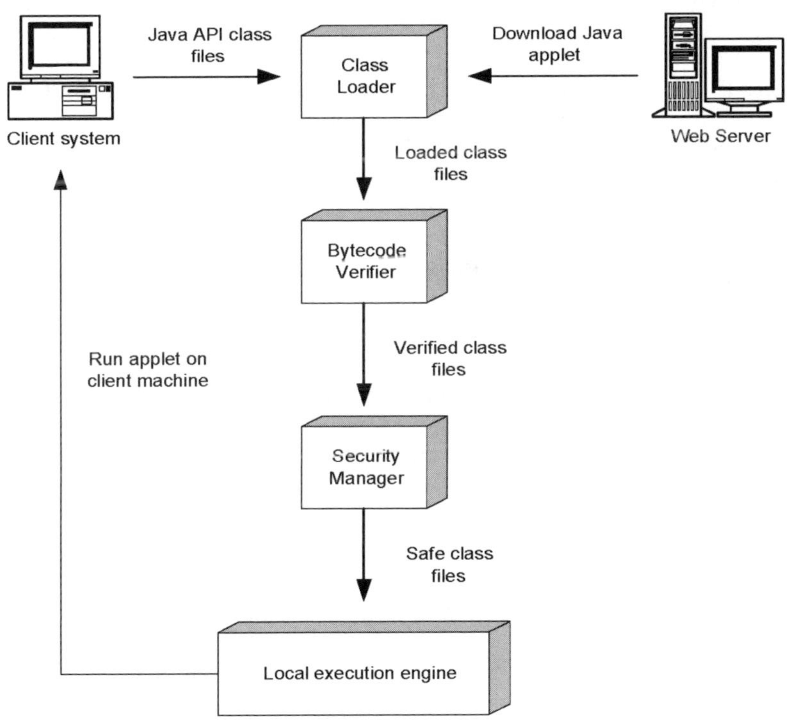

Figurc 4.9 Application of Java security model

The Class Loader determines when and how an applet can add classes to a running Java environment. An example of a running Java environment with a Class Loader is the Web browser, which makes use of Class Loader objects to download the class files for an applet across a network. When a user accesses a Web page containing an applet, the Web browser starts a Java application that installs a Class Loader object (called an *Applet Class Loader*) to download class files from a web server using the HTTP. When an applet is loaded across the network, the Applet Class Loader receives the binary data and instantiates it as a new class ready for execution.

When a Java source file is compiled, the result is platform-independent Java bytecode. All Java bytecode must be verified by the Bytecode Verifier before it can run, and verifying an applet's bytecode is one method to check untrusted downloaded code before it is allowed to run on the client machine. The Bytecode Verifier performs checking on a number of different levels to ensure the integrity of the program code. Some of these checks include checking the format of the bytecode fragments, and applying theorem provers to each bytecode fragment. The theorem prover helps to ensure that the bytecode does not forge pointers, violate access restrictions, or access objects using incorrect type information [29].

The third and final mechanism of the Java security model is the Security Manager. The Security Manager restricts the tasks that an applet can perform on a client machine, and implements a large portion of the entire security model. The Security Manager performs run-time checks on "dangerous" methods, in effect defining the outer boundaries of the sandbox. The restrictions placed on applets will be described in the following section.

4.4.2 Applet Security Restrictions

The Security Manager imposes certain security restrictions on applets to ensure that no "dangerous" actions can be done on the client system. These security restrictions can be very strict and therefore limit the functionality of applets in general. Security restrictions vary from browser to browser, with some browsers having stricter security policies than others. However, all downloaded applets are prevented from performing the following "dangerous" actions:

- *Network Restrictions* — make a network connection except to the originating host machine.
- *Library Loading Restrictions* — load dynamic libraries or define native methods.
- *System Property Restrictions* — access system properties.

However, an IJPhone Applet must perform these "dangerous" actions in order to function properly. This section explores these restrictions and explains how they can affect the functionality of the IJPhone Applet.

4.4.2.1 Network Restrictions

Java applets can only make network connections back to the Web server where they were loaded from. It may not listen for incoming socket connections or incoming UDP datagrams from anywhere except its originating server, and can only send datagrams back to its home server. This restriction affects the IJPhone Applet as it must be able to make network connections to another IJPhone Applet to send call set-up and voice data packets. Thus, being able to make network connections and listen for incoming socket connections and datagrams is central to the proper functioning of the IJPhone Applet.

4.4.2.2 Library Loading Restrictions

Applets can only make use of their own compiled Java code and are not allowed to load dynamic libraries like .dll files in Windows and .so files in Unix. They are also not allowed to define native methods specific to the native operating system. This prevents the applets from bypassing all the security features of Java by calling native methods that are used by the system classes to gain unauthorized access to the system. The IJPhone Applet needs to load dynamic libraries to access native methods for recording and playing back audio data. This is necessary as some of the recording/playback and compression/decompression functions must be done through native methods for better real-time performance. This is also due to the lack of available support for efficient and platform-independent audio recording methods in Java.

4.4.2.3 System Property Restrictions

Applets that are not run locally may not access system properties such as the host name and IP address. This is to prevent applets from changing any system properties like the `appletviewer.security.mode` property, and open a large security hole in the process. However, the IJPhone Applet needs to read several system properties including the host name and IP address during initialization.

4.4.2.4 Other Restrictions

In addition to the three most serious types of security restrictions mentioned above, applets are also prevented from performing the following actions:

- Read or write files on the host that is executing the applets.
- Execute any program on the host that is executing.

- Define their own class loaders.
- Define classes that belong to certain packages.

As with the other three restrictions, these additional restrictions also serve to prevent untrusted applets from performing unwanted actions on the host system. All downloaded applets must adhere to the security restrictions imposed by the Security Manager. The IJPhone Applet must overcome the restrictions to function properly and a solution must be found to circumvent these restrictions without compromising the security of the client's system.

To achieve this, some Web-based real-time voice systems such as the Yahoo! Voice Chat [34] makes use of ActiveX [28] to deliver their Java voice chat applet. This is done by first signing the applet and embedding it as an ActiveX control in the Web page. When a user accesses the Web page for the first time, the applet is downloaded and installed on the user's machine with the user's consent. However, this method is no different from using ActiveX directly, as the applet can only be run from one type of browser (Microsoft IE). This also defeats the purpose of writing the voice software as a platform-independent Java applet. To support the IJPhone Applet to run on Netscape Navigator as well as Microsoft IE, two possible solutions are proposed in the following sections.

4.4.3 Overcoming Security Restrictions

Java applets offer the convenience of browser-based downloading and execution, but are restricted by Java's sandbox security model. The IJPhone Applet needs to get around some of these restrictions to function properly and hence some solutions must be found to achieve this. Applets can be classified into two types — local applets and downloaded applets. Local applets are applets that are installed on the client's file system and can be run without many of the security restrictions. Downloaded applets, on the other hand, are downloaded across the Internet from a Web server and are subjected to all the applet security restrictions.

Different browsers implement different levels of security for Java applets. Netscape Navigator (Netscape) has a very strict security model, while Microsoft's IE supports several security models from "Low" (no restrictions) to "High" (will not download and run applets at all). In addition, IE allows digitally signed classes to have fewer restrictions. However, IE and Netscape relax their security policy for local applets (Web pages that are loaded with a "file://" type URL). If a file was loaded with a "http://" type URL, the applet will still be under the full scrutiny of the security manager, even if the file is stored on the local drive [35].

Table 4.1 shows the difference in the security restrictions between local and downloaded applets when viewed through Netscape Navigator. The IJPhone Applet is considered as a downloaded applet, which has stricter security restrictions as compared to a local applet [36]. The focus of this section is on overcoming the security restrictions of downloaded applets. Two proposed solutions are to use a customized security manager and authentication through code signing.

Table 4.1 Security restrictions of local and downloaded applets

Action performed	Local applets (loaded by file://)	Downloaded applets (loaded by http://)
Read system property user.name	Yes	No
Connect to port on remote client	No	No
Connect to port on third host	No	No
Load dynamic library	Yes	No

4.4.3.1 Customized Security Manager

A security manager is any Java class that is designed to extend the `java.lang.SecurityManager` class. As security managers are written in Java, they can be customized to establish a custom security policy for a Java applet or application. A browser can only have one security manager, and this is only established once, when the browser is started. The security manager remains loaded as long as the browser is running and cannot be replaced, overloaded, overridden or extended. Applets are also not allowed to create or reference their own security managers.

Java enforces its security restrictions on applets by asking the security manager for permission to take any potentially unsafe action. For each potentially unsafe action, there is a corresponding method in the security manager that determines whether the action is allowed by the sandbox. Table 4.2 shows the various methods in the Security Manager that check for any unsafe actions.

Table 4.2 Examples of Security Manager methods

Security manager method	Type of securityrestriction	Security restriction checked for
`checkAccept(String` *`host`*`, int` *`port`*`)`	Network	Accepting a socket connection from a specified host and port
`checkConnect(String`	Network	Opening a socket

Security manager method	Type of securityrestriction	Security restriction checked for
`host, int port)`		connection to a specified host and port
`checkListen(int port)`	Network	Listening for socket connections on a specified port
`checkLink(String library)`	Library loading	Loading a dynamic library that contains native methods
`checkPropertyAccess(String key)`	System property	Accessing or modifying system properties

A "check" method of the Security Manager returns a security exception if the action checked for is forbidden, and returns normally if it is permitted. When a Java method is about to perform a potentially unsafe action, it first checks whether a security manager has been installed. If one is not installed, the method goes ahead and performs the action. If one is installed, the Java API calls the corresponding "check" method in the Security Manager. If the action is forbidden, a security exception will be raised and the action will never be executed. If the action is allowed, it will be executed normally.

One solution to overcome the security restrictions imposed on the IJPhone Applet would be to create and install a customized security manager over the existing Security Manager on the client's browser. The customized security manager overrides the `checkAccept()`, `checkConnect()`, `check Listen()`, `checkLink()` and `checkPropertyAccess()` methods in the `java.lang.SecurityManager` class to allow the IJPhone Applet to access network resources, load dynamic libraries and access system properties.

4.4.3.2 Code Signing

An alternative to overcoming the security restrictions on the IJPhone Applet is through code signing. Code signing is a process where a Java applet is signed by the developer using a digital ID that could not be forged. The user that downloads the applet must specify that he/she trusts applets signed by the developer's digital ID, and this allows the "trusted" applet to override the security restrictions imposed on normal untrusted applets. A digital ID is like a signature of the developer or company, and must be verified by a CA such as Verisign [37]. A digital ID consists of two parts: a public certificate and a private key. The private key is used to sign the applet code, and the public certificate is used by users to verify that the applet was signed with the private key. Netscape and IE use different methods of code signing and authentication to give more privileges to the downloaded applets [38].

4.4.3.2.1 Netscape

The method for overcoming the security restrictions in Netscape makes use of the Netscape Capabilities [39] library, which adds facilities to refine the control provided by the Java Security Manager. The Capabilities library validates the digital signatures within a Java archive (Jar), as well as maintain a list of the kinds of accesses the user decides to allow (or disallow) for downloaded Java applets.

All access control in Netscape is a decision on who is allowed to do what. In the Capabilities security model, a Principal represents the "who" and a Target represents the "what". Each Principal contains an associated set of Privileges that represent the authorization for it to access a particular Target. When considered from the perspective of the Java Security Manager, the Principal would be the IJPhone Applet, the Target would be the required system resources, and the Privileges would be the applet security restrictions. The PrivilegeManager class within the Capabilities library keeps track of which Principals are allowed to access which targets at any given time.

The method for requesting additional Privileges for the IJPhone Applet in Netscape requires the following steps. Firstly, additional code must be included in the `init()` method of the IJPhone Applet to request Privileges from the PrivilegeManager class to perform any unsafe actions. Figure 4.10 shows part of the initialization code of the IJPhone Applet that can perform this task.

```
import java.awt.*;
import java.applet.*;
import netscape.security.PrivilegeManager;

public class IJPhone extends Applet {
   boolean hasPrivilege = false;

   // initialize Netscape security

   public void init() {
       try {
          // request Netscape 3.0 plugin
capabilities

PrivilegeManager.enablePrivilege("30Capabilities");
          hasPrivilege = true;
       } catch
(netscape.security.ForbiddenTargetException e) {
          System.out.println("Failed. Permission
denied by user.");
          e.printStackTrace();
       }

       if(hasPrivilege == false) {
          System.out.println("Necessary privileges
have not been granted. Aborting.");
          System.exit(-1);
       }

       // if we get this far, then we know the
privilege
       // is available and has been enabled. We can
activate it now.


PrivilegeManager.enablePrivilege("30Capabilities");
   }
}
```

Figure 4.10 Additional code to request Netscape Privileges

After the code has been added into the IJPhone Applet, it is then archived into a JAR file format and digitally signed using a digital certificate. The signed JAR file containing the IJPhone Applet is then embedded into the IJPhone Web Page using the Hypertext Markup Language (HTML) code shown in Figure 4.11.

```
<APPLET NAME="Internet Java Phone Applet"
        CODE="IJPhone.class" ARCHIVE="IJPhone.jar"
        WIDTH=400
        HEIGHT=50>
</APPLET>
```

Figure 4.11 HTML code to embed IJPhone Applet for Netscape

When the user downloads the IJPhone Applet, it is constrained by the sandbox and all security restrictions still apply. When the IJPhone Applet requests permission to perform an unsafe action, Netscape displays a "Java Security" dialog whether the user wants to grant or deny the applet from performing the action. The "Java Security" dialog is similar to the one shown in Figure 4.12. If the user clicks "Grant", the applet may perform the unsafe action. If the user clicks "Deny", the request fails but the applet continues execution entirely within the sandbox. Clicking on the "Certificate" button displays the digital certificate that was used to sign the applet.

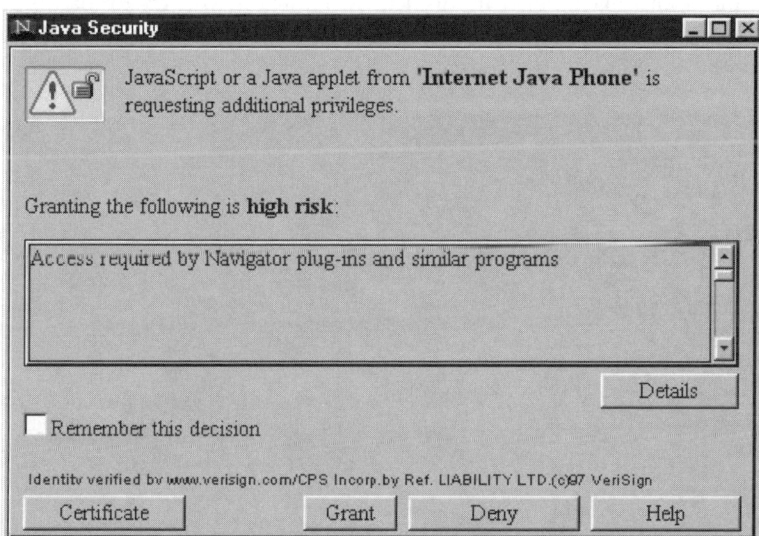

Figure 4.12 Java Security Warning dialog in Netscape

4.4.3.2.2 Internet Explorer

The method for IE is generally simpler than for Netscape. For IE, the following steps are required to remove security restrictions on the IJPhone Applet. A Microsoft archive file (IJPhone.cab) must first be created with the compiled class files of the IJPhone Applet. The archive file is then digitally signed using a signing tool.

The signed archive file containing the IJPhone Applet is then embedded into the IJPhone Web Page using the HTML code shown in Figure 4.13.

```
<APPLET NAME="Internet Java Phone Applet"
        CODE="IJPhone.class"
        WIDTH=400
        HEIGHT=50>
<PARAM NAME="CABBASE" VALUE="IJPhone.cab">
</APPLET>
```

Figure 4.13 HTML code to embed IJPhone Applet for IE

When the user downloads the IJPhone Applet, IE will display a "Security Warning" dialog asking whether the user wants to download and install the IJPhone Applet. The "Security Warning" dialog is similar to the example shown in Figure 4.14. If the user answers yes, the applet will run without security restrictions on the user's machine. If the user answers no, IE will try to load the applet using individual class files. If it succeeds, the IJPhone Applet is run entirely within the sandbox.

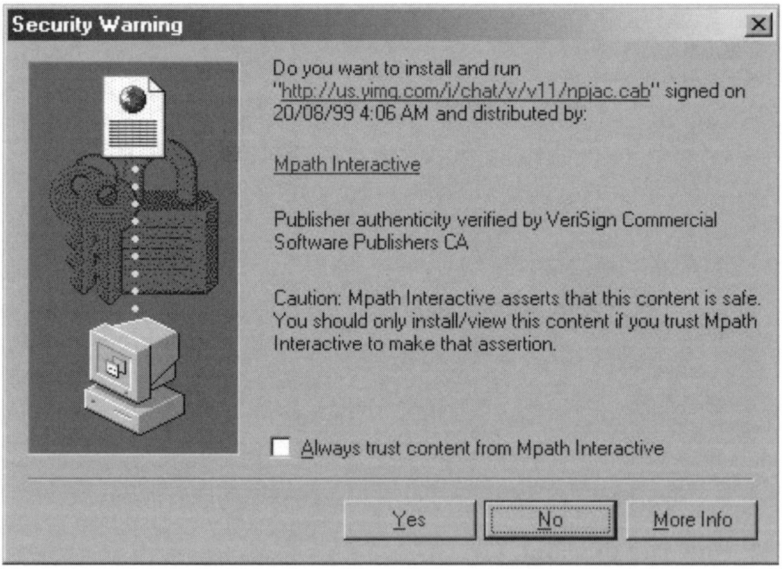

Figure 4.14 Security Warning dialog in IE

4.4.3.2.3 Dual Browser Support

Netscape and IE employ different procedures for signing applets. For Netscape, the IJPhone Applet must be signed with a Netscape Object Signing ID, and all the files packed into a .jar archive file for distribution over the Internet. For IE, all the files of the IJPhone Applet must be wrapped into a .cab archive and then signed with a Microsoft Authenticode ID. Although Netscape and IE employ different code signing methods, it is possible to embed both types of IJPhone Applet archives in a web page such that each browser will select the archive it understands and to execute it accordingly. This is done by combining the HTML code for Netscape and IE as shown in Figure 4.15.

```
<APPLET NAME="Internet Java Phone Applet"
        CODE="IJPhone.class" ARCHIVE="IJPhone.jar"
        WIDTH=400
        HEIGHT=50>
<PARAM NAME="CABBASE" VALUE="IJPhone.cab">
</APPLET>
```

Figure 4.15 HTML code to embed IJPhone Applet for both Netscape and IE

The customized security manager removes the security restrictions of the IJPhone Applet by overriding the browser's default security manager. However, designing and implementing a security manager is a complicated process and any small error in programming can expose the client system to attacks by other malicious applets. In addition, the user must install the customized security on his/her Web browser before downloading the IJPhone Applet. This additional effort of installation makes it more inconvenient for the user to use the software, and at the same time does not guarantee the security of the client system.

The code-signing method offers a better alternative in the form of digital signatures and authentication. By allowing the user to determine whether or not to trust the downloaded IJPhone Applet, the IJPhone Applet is either allowed full access to the client system or is only allowed to run entirely within the sandbox. The use of digital signatures means that users do not need to install any code in advance and only need to grant access to the IJPhone Applet after it has been completely downloaded. There is no additional overhead involved on the part of the user except that the user must grant permission to the IJPhone Applet to allow it to run outside the sandbox. In addition, the code-signing method is performed after the IJPhone Applet has been completely implemented, and it does not affect the overall architecture of the system. Hence, the code-signing method is the preferred way to overcome the security restrictions imposed on the IJPhone Applet.

4.5 Biometrics for Identity Authentication — Multi-view Facial Analysis

Traditional methods of person recognition are inadequate. Large sums of money have changed hands on the basis of a signature, identity cards and/or passwords, all of which are unreliable. An imposter can fake a signature, identity cards can be forged and passwords can be lost, stolen or forgotten. There is an increasing need for automatic person recognition systems that are fast, accurate and user-friendly.

Biometrics [40, 41] is an area of research gaining importance in recent years. Biometrics relates to a person's physiological features (fingerprints, face) and behaviours (speech), and these can be used to distinguish one person from others. In fact, fingerprinting technology has been used in forensics by law enforcement agencies for many years.

Biometric systems, which use people's physiological characteristics for authentication, have become increasingly popular to counter fraud and other misdeeds. The purpose of authentication is to determine if a person is who he/she claims to be, which is a binary decision. This section describes a face recognition system that uses a novel distance measure for authentication [42]. This provides much better results and tolerance than other methods. Face recognition is chosen because it is one of the least intrusive biometric approaches. This is an important consideration if the system is to be adopted widely by voluntary users. Possible applications include access control to sensitive databases, installations, and so on and online trading. Although the goal is to optimize the performance of single-modal operation, this proposal is generic enough for combining other modes of operation (e.g. speech analysis) if desired.

In this field of research, the primary performance measure is the trade-off between false acceptance rate (FAR) and false rejection rate (FRR). This trade-off then leads to the consideration of an equal error rate (EER) line. Typically, we are interested in plotting FRR against FAR to compare the performance of different systems and/or system parameters. The curves thus obtained are known as *receiver operating characteristic* (ROC) curves. EER is then the line given by FRR = FAR in a ROC plot.

4.5.1 The Need for an Effective Distance Measure

The fundamental question that authentication attempts to answer is "is the person claming to be person *X* really person *X*?" In essence, each user who voluntarily takes part must have his/her biometrics features stored in a database during an enrolment process. Subsequently, when the user attempts to gain access to protected resources, he/she must be authenticated. During authentication, the system must compare the stored features with those newly captured features. For speech-related authentication, the features will be extracted from a speech segment. In the case of facial analysis, the features are extracted from two-dimensional (2D) images of each user's face. The authentication process then boils down to a matching process between the stored images and the newly captured image(s).

The need for a quantitative analysis of a match between two images leads to the consideration of a distance measure. For example, suppose an image A is represented by a set of feature vectors (nodes) N_A (model points) and image B is represented by another set N_B (test points). Then, matching images A and B amounts to determining whether the distances between the respective nodes in N_A and N_B are sufficiently close. An authentication system will then make a binary decision based on this overall proximity. An effective distance measure is therefore critically important in determining the performance of a face recognizer for authentication.

The Hausdorff distance is a popular distance measure well suited to establishing similarity between two sets of points, for example, [43]. Unlike most other methods that rely on a rigid point-to-point relationship between model and test points, Hausdorff distance can be determined without such explicit relationship. This makes Hausdorff distance particularly useful for comparing non-rigid objects such as faces. In fact, some commercial products have adopted the use of Hausdorff distance, for example, BioID [44] uses it to establish the face location and eye location.

Mathematically, if we have two finite sets of feature points $M = \{m_1, m_2, ..., m_k\}$ (representing a model) and $T = \{t_1, t_2, ..., t_n\}$ (representing a test image), the Hausdorff distance is defined as

$$H(M,T) = \max(h(M,T), h(T,M)) \tag{4.1}$$

where

$$h(M,T) = \max_{m_i \in M} \min_{t_j \in T} \left\| m_i - t_j \right\| \tag{4.2}$$

and $\| m_i - t_j \|$ denotes the Euclidean norm on the points of M and T. The function $h(M,T)$ is called the *directed Hausdorff distance* from M to T. It identifies the point $m_i \in M$ that is the farthest from any point of T and measures the distance from m_i to its nearest neighbour in T. The Hausdorff distance $H(M,T)$ is the maximum of $h(M,T)$ and $h(T,M)$. Thus, it measures the degree of mismatch between two sets by measuring the distance of the point of M that is farthest from any point of T and vice versa.

However, the major drawback of this traditional definition of the Hausdorff distance is that it is very sensitive to perturbations caused by even a few outlying points that may have resulted from imperfection in the image segmentation process. A modified Hausdorff distance (MHD) has been found to be less susceptible to outlying points [45]. Unfortunately, all reported variants of the Hausdorff distance do not address the issue of unequal contribution of each point. In practice, the prominence of each point in representing a facial image is likely to be different. In this research, we apply a new variant of Hausdorff distance known as *significance-based multi-view Hausdorff distance* (SMVHD). This considerably improves the robustness of the authentication process even with the introduction of non-rigid distortions to the facial image, for example, when a person speaks.

4.5.2 The Significance-based Multi-view Hausdorff Distance

Traditional methods that assign the same weight to all points in N_A and N_B clearly lead to sub-optimal results because different feature points contribute differently towards an overall description of a facial image. In addition, we attempt to fuse together multiple view images to obtain better performance than frontal-only approaches without the extra expense incurred in a multi-modal system (such as when integrating facial analysis with speech analysis). It should be stressed that by attempting to optimize a single-modal operation (facial analysis), we do not preclude the fusion of this technique with other techniques (e.g. speech analysis). In fact, the robustness against non-rigid distortions (e.g. facial motion due to speaking) would make our technique well suited to multi-modal operations if desired.

SMVHD differs from other variants of Hausdorff distance (HD) in two ways. First, there is a measure of significance associated with each point. Multiple views of the same non-rigid object (the face in this case) taken from different viewpoints are fused together. More specifically, suppose $M^1 = \{m_1^1, m_2^1, \cdots, m_{p_1}^1\}$, $M^2 = \{m_1^2, m_2^2, \cdots, m_{p_2}^2\}, \ldots, M^n = \{m_1^n, m_2^n, \cdots, m_{p_n}^n\}$ are n point *sets* representing the features in the n model views of object M and $T^1 = \{t_1^1, t_2^1, \cdots, t_{q_1}^1\}$, $T^2 = \{t_1^2, t_2^2, \cdots, t_{q_2}^2\}, \ldots, T^n = \{t_1^n, t_2^n, \cdots, t_{q_n}^n\}$ are the corresponding n point *sets* representing the features in the n views of the test object T from the same viewpoints as in the model M. In this formulation, $p_1, p_2, \ldots, p_n$ and $q_1, q_2, \ldots, q_n$ are the point numbers in the model and the test views, respectively, which are used for indexing. Thus, the ith point in the kth viewpoint $t_i^k \in T^k$ must only be allowed to match with the points in M^k, the kth view of M. The SMVHD between M and T is then defined as

$$H_{SMVHD}(M,T) = \max(h_{SMVHD}(M,T), h_{SMVHD}(T,M)) \tag{4.3}$$

The directed SMVHD from *M* to *T* and from *T* to *M* are in turn defined as

$$h_{SMVHD}(M,T) = \frac{1}{\sum_{k=1}^{n} \sum_{m_i^k \in M^k} Sig_{m_i^k t_j^k}} \sum_{k=1}^{n} \sum_{m_i^k \in M^k} (Sig_{m_i^k t_j^k} \bullet \min_{t_j^k \in T^k} \left\| m_i^k - t_j^k \right\|) \tag{4.4}$$

$$h_{SMVHD}(T,M) = \frac{1}{\sum_{k=1}^{n} \sum_{t_i^k \in T^k} Sig_{t_i^k m_j^k}} \sum_{k=1}^{n} \sum_{t_i^k \in T^k} (Sig_{t_i^k m_j^k} \bullet \min_{m_j^k \in M^k} \left\| t_i^k - m_j^k \right\|) \tag{4.5}$$

where $Sig_{m_i^k t_j^k} = \frac{1}{2}(Sig_{m_i^k} + Sig_{t_j^k})$ is the average significance of point m_i^k and its corresponding point t_j^k. Compared to other variants of Hausdorff distance where the contributions from all the points are equal, every $\min_{t_j^k \in T^k} \left\| m_i^k - t_j^k \right\|$ (i.e. the distance of a matched pair) in SMVHD is weighted by the average significance of m_i^k and t_j^k because its contribution to $h_{SMVHD}(M,T)$ is assumed to be proportional to the significances of the two matched points. The same property applies to $h_{SMVHD}(T,M)$. In fact, SMVHD is symmetric with respect to *M* and *T*. So, matching *M* to *T* will yield the same result as matching *T* to *M*. In addition, multiple points in *M* may be matched to a single point in *T* and vice versa. This introduces tolerance to non-rigid distortion into the distance measure.

4.5.3 An Experimental System

We have developed an experimental access control system based on the fusion of multiple views of facial images using the SMVHD discussed above. One possible application scenario of the system is depicted in Figure 4.16, where a stationary person stands at a designated position to have multiple facial images taken.

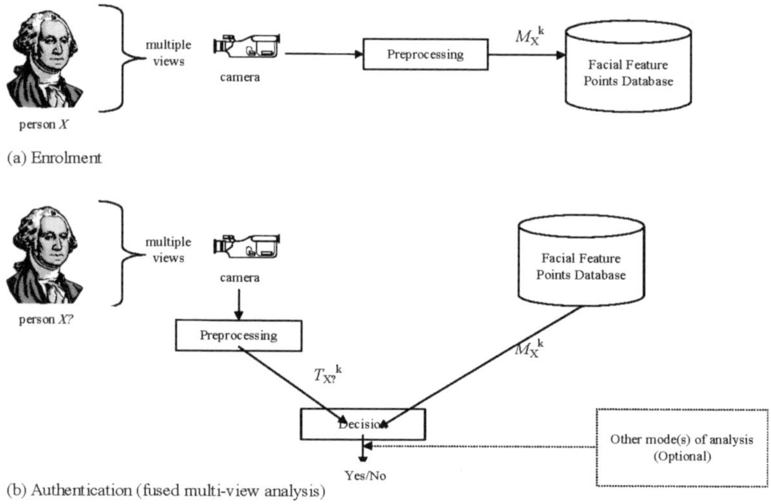

Figure 4.16 Overview of the experimental system. (a) Enrolment. (b) Authentification (fused multi-view analysis)

The multiple views can be obtained either by installing multiple fixed cameras at the desired locations (three in our case), or to have a single camera that moves on a track. With the availability or modification of suitable hardware components, the system could also be used for other authentication purposes such as online business transactions. During enrolment as shown in Figure 4.16(a), three views (frontal, ¾ and profile) are captured as still images. Pre-processing then isolates the facial area in order to reduce the effect of hair and clothing. For the profile image, pre-processing entails nose tip and chin detection, followed by normalization, alignment and cropping of the facial area. For the other images, the pre-processing procedure is the same except that eye detection is performed instead of nose tip and chin detection [46]. The extracted model points are stored in a database as M_X^k for person X with views $k = 1,2,3$.

During authentication as shown in Figure 4.16(b), someone who claims to be person X has his/her multiple views captured as still images and pre-processed with the same procedure as in the enrolment phase. This results in $T_{X?}^k$ for someone claiming to be person X. Based on SMVHD matching with the stored model for person X (M_X^k), a decision is made on whether this is in fact person X. For generality, our system also allows a final decision to be made by combining results from other modes of analysis (e.g. speech). However, this is not currently implemented as we are interested in optimizing the performance of multi-view facial analysis, which essentially is a single-modal operation.

4.5.4 System Performance

We have used both multi-view facial images where we captured ourselves along with the images from the University of Stirling database. The latter contains 311 images of 35 persons [47]. In particular, there are 31 complete sets of images (16 females and 15 males), each of which is complete with three poses (frontal, ¾ and profile views) and three expressions (neutral, smiling and speaking).

In the case of the images we captured, we designed the experiment to simulate real-life situation for access control as far as possible. This meant introducing a delay between images captured for the enrolment process and for authentication. In addition, we randomly altered the illumination to simulate a typical noisy real-life environment. In particular, we conducted our experiments in a laboratory with windows on one side and we varied the mix of ambient sunlight with room lights (fluorescent tubes).

The experiments were conducted based on a "leave-one-out and rotation" scheme. In particular, we labelled each person as an "imposter" in turn, with the others acting as "clients". The imposter's role was to attempt to be identified as one of the 20 clients. The number of times that each imposter was able to gain access under someone else's identity would lead to FAR. The clients, on the other hand, were asked to gain access under their own identity, which would result in FRR. In particular, FAR is given by the number of imposter acceptances divided by the number of imposter claims; FRR is obtained by dividing the number of client rejections divided by the number of client claims. The experimental protocol is described diagrammatically in Figure 4.17. Figure 4.18 gives a ROC plot for our experimental system. The ROC plot shows that this system performs very well compared to other similar systems.

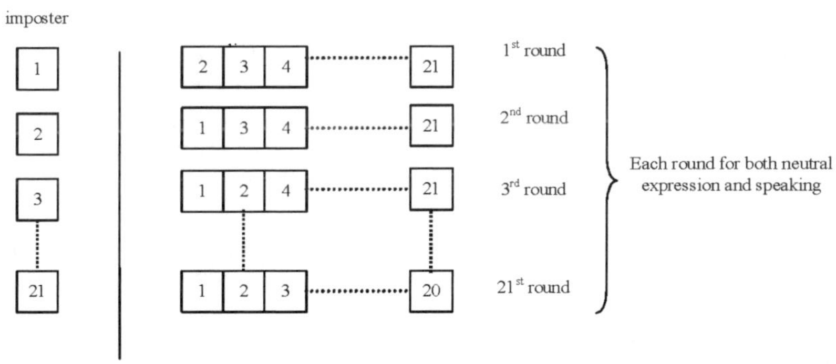

Figure 4.17 Experimental protocol — a "leave-one-out and rotation" scheme

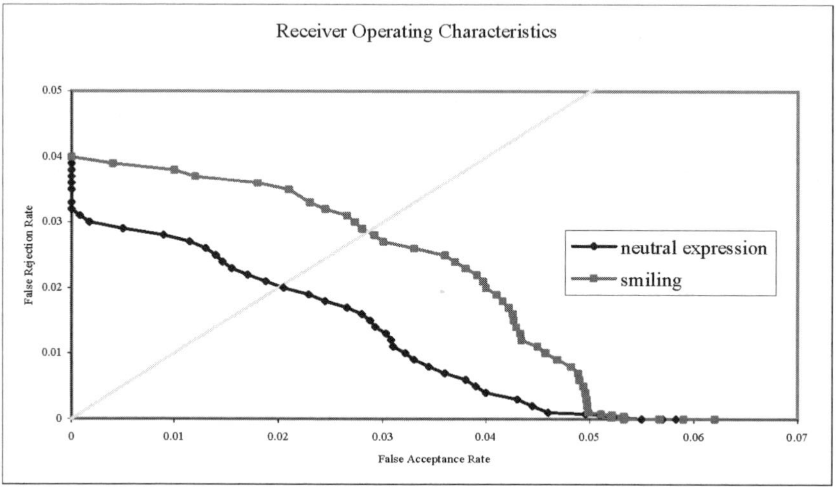

Figure 4.18 ROC plot for the experimental system

References, Links and Bibliography

[1] http://www.w3.org/Security/faq/www-security-faq.html, 2004.

[2] http://www.teu.ac.jp/siit/~tominaga/anticrack, 2004.

[3] B. Schneier, "*Applied Cryptography: Protocols, Algorithms, and Source Code in C*", 2nd Edition, John Wiley & Sons, New York, 1996.

[4] R.M. Needham and M.D. Schroeder, "Using encryption for authentication in large networks of computers", *Communications of the ACM*, Vol. 21, No. 12, pp. 993–999, 1978.

[5] http://www.rsa.com/rsalabs/newfaq, 2004.

[6] http://www.certicom.com/ecc/wecc2.htm, 2004.

[7] M. Abadi and R. Needham, *Prudent Engineering Practice for Cryptographic Protocols*, SRC Research Report 125, Systems Research Centre, Digital Equipment Corporation, 1994.

[8] http://www.real.com/devzone/library/whitepapers/security.html, 2004.

[9] B. Adida, "Weaving the web: identity crisis on the web'', *IEEE Internet Computing*, Vol. 1, pp. 91–93, Sept.-Oct., 1997.

[10] K. Pleas, "*Certificates, Keys, and Security*'', published as PC Tech Feature in the 4/20/99 issue of PC Magazine. http://www.zdnet.com/pcmag/pctech/content/18/08/tf1808.001.html, 2004.

[11] http://www.rsa.com/standards/, 2004.

[12] B. Adida, "Weaving the web: securing the web'', *IEEE Internet Computing*, Vol. 1, pp. 91–93, May-June, 1997.

[13] Netscape, DevEdge Online Document, "*Secure Sockets Layer*'', http://developer.netscape.com/tech/security/, 2004.

[14] http://www.rsasecurity.com/products/, 2004.

[15] F.M. Avolio, "A multi-dimensional approach to internet security'', *IEEE/ACM Transactions on Networking*, Vol. 1, April/May, pp. 15–22, 1998.

[16] A. Menezes, P. Oorschot and S. Vanstone, "*Handbook of Applied Cryptography*'', CRC Press, http://www.cacr.math.uwaterloo.ca/hac/about/chap8.pdf

[17] RSA, "*Public-key Encryption*'', Data Security Inc, http://www.rsasecurity.com/rsalabs/faq/2-1-1.html

[18] RSA, "*Data Encryption Standard (DES)*'', Data Security Inc, http://www.rsasecurity.com/rsalabs/faq/3-2-1.html>

[19] RSA, "*RSA Encryption*'', Data Security Inc, http://www.rsasecurity.com/rsalabs/faq/3-2-1.html

[20] PGP International, "*Pretty Good Privacy (PGP)*'', http://www.pgpi.org, 2004.

[21] http://home.netscape.com/products /security/index.html, 2004.

[22] http://www.microsoft.com/security, 2004.

[23] http://www.orasis.com/ORASIS/wbd/security.htm, 2004.

[24] http://www.openmarket.com, 2004.

[25] http://www.eit.com/creations/s-http, 2004.

[26] http://www.nai.com/products/security/freeware.asp, 2004.

[27] http://www-leland.stanford.edu/group/DCE /Gaurav/sec_doc.html, 2004.

[28] S. Kaufman, J. Perkins Jr. and D. Fleet, "*Teach Yourself ActiveX Programming in 21 Days*", Sams.Net Publishing, Indianapolis, IN, 1996.

[29] E. Felten and G. McGraw, "*Understanding the Keys to Java Security – the Sandbox and Authentication*", JavaWorld, May 1997.

[30] K.V. Chin, S.C. Hui and S. Foo, "Enhancing the quality of internet voice communication for internet telephony systems'', *Journal of Network and Computer Applications*, Vol. 21, pp. 203–218, 1998.

[31] S. Foo and S.C. Hui, "A framework for evaluating Internet telephony systems'', *Internet Research*, Vol. 8, No. 1, pp. 14–25, 1998.

[32] A. Silberschatz and P.B. Galvin, "*Operating System Concepts*", Fifth Edition, Addison-Wesley, Reading, MA, 1998.

[33] B. Venners, "*Java's Security Archtecture*", JavaWorld, August, 1997, http://www.javaworld.com/javaworld/jw-08-1997/jw-08-hood_p.html

[34] http://chat.yahoo.com, 2004.

[35] M. Wutka, "*Applet Security Restrictions*", EarthWeb, 1997, http://www.itlibrary.com/library/078970935x/ch3.htm

[36] http://java.sun.com/sfaq, 2004.

[37] http://www.verisign.com, 2004.

[38] D. Griscom, "*Code Signing for Java Applets*", 1998, http://www.suitable.com/Doc_CodeSigning.shtml

[39] Netscape Corporation, "*Introduction to the Capabilities Classes*", http://developer.netscape.com/docs/manuals/signedobj/capabilities/index.html, 2004.

[40] S. Pankanti, R.M. Bolle and A. Jain, "Biometrics: the future of identification'', *IEEE Computer*, Vol. 33, No. 2, pp. 46–49, Feb., 2000.

[41] S. Liu and M. Silverman, "A practical guide to biometric security technology'', *IEEE Computer Society IT Professional*, Vol. 3, No. 1, pp. 27–32, Jan/Feb., 2001.
[42] Y. Gao, S.C. Hui and A.C.M. Fong, "A multi-view facial analysis technique for identity authentication'', *IEEE Pervasive Computing*, Vol. 2, No. 1, pp. 38–45, 2003.
[43] D.P. Huttenlocher, G.A. Klanderman and W.J. Rucklidge, "Comparing images using the Hausdorff distance'', *IEEE Transactions on Pattern Analysis and Machine Intelligence*, Vol. 15, pp. 850–863, Sept., 1993.
[44] R.W. Frischolz and U. Dieckmann, "BioID: a multimodal biometric identification system'', *IEEE Computer*, Vol. 33, No. 2, pp. 64–69, Feb., 2000.
[45] M.P. Dubuisson and A.K. Jain, "A Modified Hausdorff Distance for Object Matching'', *Proceedings of the 12th International Conference on Pattern Recognition*, Jerusalem, Israel, pp. 566–568, 1994.
[46] K.M. Lam and H. Yan, "Locating and extracting the eye in human face images'', *Pattern Recognition*, Vol. 29, pp. 771–779, 1996.
[47] University of Stirling Face Database. http://pics.psych.stir.ac.uk/ or http://www.stir.ac.uk/Departments/HumanSciences/Psychology/, 2004.

CHAPTER 5

INTERNET PRIVACY

5.1 INTRODUCTION

This chapter deals with the issue of Internet privacy, in particular, effective Web content filtering techniques. The open and unregulated nature of the Internet makes it a possible forum for some individuals to abuse their rights of free expression. Unsuspecting users can easily be led to objectionable Web contents just by making a simple typing mistake when specifying a URL, for example. These objectionable Web contents, such as hate messages, pornography, violence, gambling, and so on, can be very harmful, particularly for youngsters. In addition, spam can also be quite irritating for users.

For companies and government agencies, the management may also want to minimize the time employees spend on surfing the Internet for non-work-related Web contents. This should lead to an improvement in productivity. All these point to the need for an effective way to screen (or filter) Web contents that are allowed to reach a particular terminal.

We present the latest techniques in combating the rampant rise in the proliferation of objectionable Web contents. For many parents, effective tools for screening out objectionable Web contents cannot be introduced fast enough. Our survey of existing techniques and systems has led to the conclusion that most existing techniques are inadequate. They are not accurate or fast enough, and most lack the learning capabilities to adapt to the volatile nature of Web contents. We have therefore set out to develop our own intelligent Web content filtering system that performs two tasks: offline analysis and online filtering. It turns out that this is a very good approach when combined with other appropriate techniques, such as the use of artificial intelligence for learning to adapt to new problems. In particular, the following topics will be presented:

- Survey of Web content filtering technique.
- An effective Web content filtering system.

Multimedia Engineering A. C. M. Fong & S. C. Hui

5.2 WEB CONTENT FILTERING METHODS AND TOOLS — A SURVEY

The Web has become an extremely popular communications medium due to its wide geographical coverage and continuous availability. In addition, it is very easy for anyone to put up information on the Web to reach a wide audience that continues to grow in size. However, the self-regulating nature of the Web community, coupled with the ease of making information available on the Web, has given the opportunity for some individuals to abuse their expression of freedom by putting up objectionable materials on the Web. These include violence, gambling, drugs and pornographic materials. With many children who already have access to the Internet, it is of urgent importance to provide effective Web content filtering to protect children and other unsuspecting users from the adverse effects of such harmful materials. Effective Web content filtering tools are therefore needed.

Web content filtering would also benefit companies that want to reduce the overheads associated with their employees' non-work-related access to the Internet. Unauthorized access to the Internet not only adds costs to the company in terms of increased data traffic charges, but also loss of productivity by the offending employees. Sometimes, when valuable bandwidth resources are used for unauthorized access to the Web, other employees who have a legitimate need to access the Internet might also be affected.

This section presents a survey of methods and systems that provide Web filtering to users. We use the filtering of pornographic materials as a case study because these are among the most prolific and harmful. Our survey reveals that there are significant shortcomings in these solutions. It is necessary to analyse and characterize the offending Web pages in order to develop effective filtering techniques against them. In the next section, we present the results of our study of the distinguishing features of one particular type of offending Web materials. With this knowledge, we have developed an intelligent classification engine for effective Web filtering.

5.2.1 Current Methods

Four major approaches have been developed for Web content filtering. They are Platform for Internet Content Selection (PICS) , URL blocking, keyword filtering and intelligent content analysis.

5.2.1.1 PICS

PICS [1] is a set of specifications for content rating systems. It enables web publishers to associate labels or meta data with web pages to limit certain web contents to target audiences. There are two popular PICS content rating systems: RSACi [2] and SafeSurf [3]. Created by Recreational Software Advisory Council,

RSACi uses four categories and a number for each category indicating the degree of potentially offensive contents. The four categories include harsh language, nudity, sex and violence. Each number can range from 0 (nil) to 4 (highest). SafeSurf is a much more detailed content rating system which identifies the appropriateness of websites for specific age groups, in addition to the 11 categories to describe the potential offensiveness of the Web contents. Each category has nine levels, from 1 (nil) to 9 (highest).

Currently, both Microsoft Internet Explorer and Netscape Navigator, as well as several web filtering systems offer PICS support and are capable of filtering web pages according to the embedded PICS rating labels. However, PICS is a voluntary self-labeling system and each web content publisher is totally responsible for rating the contents. Consequently, a web filtering system should use PICS only as a supplementary filtering approach.

5.2.1.2 URL Blocking

This is a technique that restricts or allows access by comparing the URL (and equivalent IP addresses) of the requested web page with those in a stored list. Two types of lists can be maintained: black list and white list. A black list contains URLs of objectionable websites to be blocked, while a white list contains URLs of permissible websites. Most of the current web filtering systems that employ URL blocking use the black list approach.

The chief advantage of this approach is speed and efficiency. A filtering decision can be made by matching the URL string of the requested web page with those in the URL list even before a network connection to the remote web server is made. However, this approach requires the implementation of a URL list and only those sites with their URLs in the list can be identified. Also, unless the list is updated constantly, the accuracy of the system will fall over time due to the explosive growth of new websites. Most of the available web filtering systems utilizing URL blocking employ a large team of human reviewers to actively search for objectionable websites to be added into the black list, which is then made available for downloading as an update to the local copy of the list. This is not only time-consuming, but also resource intensive.

Despite the drawback of this approach, its fast and efficient operation is greatly desirable in a web filtering system. Using sophisticated content analysis techniques in the classification process, the content nature of a web page can be first identified. If it is determined to contain objectionable material, the URL of the Web page can be added to the black list. Later, if the Web page is accessed, a filtering decision can be made immediately by matching the URL. By dynamically updating the black list, speed and efficiency can be achieved while accuracy is maintained, provided the content analysis performed in the classification process is accurate in determining the content nature of the Websites.

5.2.1.3 Keyword Filtering

Keyword filtering is an intuitively simple approach that blocks access to websites based on the occurrence of offensive words and phrases in the Web contents. When a Web page has been successfully retrieved from the remote Web server, every word or phrase on the Web page is compared against those in a keyword dictionary of prohibited words and phrases. Blocking occurs if the number of matches has reached a pre-defined threshold.

It is a fast and simple content analysis method that can quickly determine if a web page contains potentially harmful material. However, the approach is well known for over-blocking, that is, blocking a large number of websites that do not contain objectionable context. As it carries out its filtering by matching keywords or key-phrases such as "sex" and "breast", websites about sexual harassment or health information concerning breast cancer, or even a home page of a person named *Sexton* can be accidentally blocked. Although the dictionary of objectionable words and phrases does not require frequent updates, the high over-blocking rate greatly jeopardizes the filtering capability of the Web filtering system and is often not acceptable. However, a web filtering system can use this approach to decide whether to further process a web page by using a more precise content analysis method that usually requires a longer processing time.

5.2.1.4 Intelligent Content Analysis

A Web filtering system can make use of intelligent content analysis approaches to perform automated classification on the Web contents. One of the interesting ways of implementing such a capability is by using artificial neural networks (NN) [4] that can learn and adapt according to the training cases fed to the networks. Such a learning and adaptation process can give semantic meaning to context-dependent words, such as "sex" which can occur frequently in both pornographic and other (e.g. health-related) Web pages. To achieve high accuracy in classification, the NN should be trained with a sufficiently large number of training exemplars, including both positive and negative cases. Inputs to the NN can be characterized from the Web pages such as the occurrence of keywords and key-phrases, and hyperlinks to other similar websites.

5.2.2 Current Systems

According to the location of system deployment, Web content filtering systems can be considered client-based or server-based. A client-based system performs Web content filtering solely on computers where it is installed without the need to consult remote servers about the nature of the Web contents that a user tries to access. Server-based systems are developed as server software providing filtering service to computers on the Local Area Network (LAN) where the systems are installed.

They screen the outgoing Web requests and analyse incoming Web pages to determine the content type and carry out blocking at the server-side before the Web pages reach the Web browser on client computers.

Table 5.1 summarizes the features of 10 popular Web content filtering systems. These are Cyber Patrol [5], Cyber Snoop [6], CYBERsitter [7], I-Gear [8], Net Nanny [9], SmartFilter [10], SurfWatch [11], WebChaperone [12], Websense [13] and X-Stop [14]. Only two systems are specifically geared towards filtering pornographic websites. As expected, none of the ten systems relies on PICS as the main technique. Six of the systems mainly rely on URL blocking while only two systems use keyword filtering as their main content filtering approaches. I-Gear incorporates both URL blocking and a proprietary dynamic content filtering technique called *Dynamic Document Review*, which looks for matches with keywords dynamically. Only WebChaperone employs content analysis as its main content filtering approach. It utilizes a unique mechanism called *Internet Content Recognition Technology* (iCRT) to dynamically evaluate each web page before it is passed to the Web browser. iCRT analyses a Web page's attributes including word count ratios, length of page, structure of page, and contextual phrases. The results are then aggregated according to the weighting of the attribute. According to the overall results obtained, WebChaperone identifies whether the Web page contains pornographic material.

Table 5.1 Feature comparison of ten popular web content filtering systems

Systems	**Location**	**Content filtering approach**				**Filtering domain**
		PICSs upport	**URL blocking**	**Key-word filtering**	**Con-tent analy-sis**	
Cyber Patrol	Client	Yes	Yes	Yes	No	General
Cyber Snoop	Client	Yes	Yes	Yes	No	General
CYBERsitter	Client	Yes	Yes	Yes[1]	No	General
I-Gear	Server	Yes	Yes	Yes[2]	No	General
Net Nanny	Client	No	Yes	Yes	No	General
SmartFilter	Server	No	Yes	No	No	General
SurfWatch	Client	Yes	Yes	Yes	No	General
WebChaper-one	Client	Yes	Yes	Yes	Yes[3]	Porno-graphic
Websense	Server	No	Yes	Yes[4]	No	General
X-Stop	Client	Yes	Yes	Yes[5]	No	Porno-graphic

1. Context-based key-phrase filtering technique
2. Employed in Dynamic Document Review and software robots
3. Content analysis carried out using iCRT
4. Employed only in web crawlers
5. Employed in MudCrawler

5.2.2.1 Performance Analysis

We have evaluated Cyber Patrol, Cyber Snoop, CYBERsitter, SurfWatch and WebChaperone according to the underlying major content filtering approaches employed so that we can gauge the accuracy of different approaches. Each of the five systems was installed on individual computers and we attempted to visit different websites while the web filtering system was active. The URLs of 200 pornographic and 300 neutral web pages were collected and used for evaluation. To isolate and measure the accuracy of the individual content filtering techniques, each system was limited only to its major approach. So, Cyber Patrol and SurfWatch were configured to use the URL blocking technique, while Cyber Snoop, CYBERsitter and WebChaperone utilized only keyword filtering, context-based key-phrase filtering, and iCRT, respectively.

Table 5.2 summarizes the results of our evaluation. The overall accuracy is obtained by averaging the correctly classified web page percentage for both pornographic and neutral web pages. For the two systems that employ URL blocking, the number of incorrectly classified neutral web pages is very small compared to those using keyword filtering. This shows that the black list compiled by the specialists has accurately excluded URLs of most neutral websites, even if the websites contain sexually explicit terms that are not used in a pornographic context. However, both systems have fairly high occurrences of incorrectly classified pornographic web pages. This highlights the problem of keeping the black list up to date.

Systems that rely on keyword filtering tend to perform well on pornographic web pages, but the percentage of incorrectly classified neutral web pages can be very high. This highlights the major shortcoming of the keyword approach. On the other hand, WebChaperone, which employs iCRT, achieves the highest overall accuracy of 91.6% among the five systems. This underlines the effectiveness of utilizing content analysis-based approaches to achieve high accuracy in web filtering systems.

Table 5.2 Performance of five popular web content filtering systems

System	Major approach	Nature of web page				Overall accuracy
		Pornographic (Total: 200)		Neutral (Total: 300)		
		Correctly classified	Incorrectly classified	Correctly classified	Incorrectly classified	
Cyber Patrol	URL blocking	163 (81.5%)	37 (18.5%)	282 (94.0%)	18 (6.0%)	87.75%
SurfWatch	URL blocking	171 (85.5%)	29 (14.5%)	287 (95.7%)	13 (4.3%)	90.6%
Cyber Snoop	Keyword filtering	187 (93.5%)	13 (6.5%)	247 (82.3%)	53 (17.7%)	87.9%
CYBER-sitter	Context-based key-phrase filtering	183 (91.5%)	17 (8.5%)	255 (85.0%)	45 (15.0%)	88.25%
WebChaperone	iCRT	177 (88.5%)	23 (11.5%)	284 (94.7%)	16 (5.3%)	91.6%

Using the results of our survey as a benchmark, we set out to develop our own intelligent Web content filtering technique, which is described in the next section.

5.3 An Effective Web Content Filtering System

Intelligent content analysis is an attempt at achieving semantic understanding of the context in which certain keywords appear. In particular, intelligent classification techniques can be used to categorize Web pages into different groups (e.g. pornographic and non-pornographic) according to the statistical occurrence of sets of features. Statistical methods such as K-Nearest Neighbour (KNN) classification [15] [16], Linear Least Squares Fit (LLSF) [17][18], Linear Discriminant Analysis (LDA) [19] [20] and Naïve Bayes (NB) probabilistic classification [21] have been introduced in this field of research. In addition, NN models [22] [23] are well suited to providing categorization on real-world data characterized by incomplete and noisy data. However, the use of statistical and NN techniques can be compute-intensive and often incur intolerable latency.

In this research, we decouple the Web page classification process from the filtering process to achieve fast and effective filtering from the users' point of view [24]. The classification process is conducted offline by NN, whose learning capabilities allow it to adapt to noisy data and acquire human-like intelligence in distinguishing the nature of a Web page by semantic understanding of the context in which keywords appear. We investigate two popular NN models that are effective classifiers: Kohonen's Self-Organizing Map (SOM) [25] [26] and Fuzzy Adaptive Resonance Theory (Fuzzy ART) [27]. This offline intelligent classification process is used to create and maintain a knowledge base of prohibited URLs without the need for human expertise or supervision. This means the online filtering process can be very fast and effective. To achieve this, we need to study the characteristics and distinguishing features of pornographic pages.

5.3.1 Analysis of the Target Web Pages

We attempt to identify the characteristics of pornographic Web pages by analysing the textual and page layout information contained in such Web pages. We also investigate the adoption rate of PICS, with an understanding that Web page classification could not totally rely on it. PICS could be used for the positive identification of pornographic Web pages. We present the results of our analysis on a sample of Web pages from 200 different pornographic Websites.

5.3.1.1 Page Layout

Like other Web pages, pornographic Web pages can be classified into two layout formats: single-frame and multi-frame. When viewing a Web page with single-frame layout in a Web browser, the browser only needs to download one HTML document from a single URL address and construct the whole Web page contents from that document. On the other hand, to view a multi-frame Web page, an HTML document containing information of other HTML documents that make up the contents of the whole Web page is downloaded first. The information includes the URL addresses of the HTML documents, as well as the position data of where to display the specific document in the browser window. According to this piece of information, the browser fetches all the necessary HTML documents and constructs the whole Web page.

Consequently, all the HTML documents used in constructing a multi-frame Web page must be treated as a single entity. This is an important consideration because when any multi-frame Web pages are encountered in the data collection and analysis process, the statistics obtained from any aspects of the Web page should be derived as a whole from the aggregation of data collected from every individual HTML document making up the whole Web page.

We have studied how widespread multi-frame Web pages are among pornographic Websites in order to determine whether it is necessary to gauge the importance of incorporating the processing capability of such Web pages in our system. We found that while an overwhelming majority of pornographic Web pages adopt the single-frame format (86%), 13% used a two-frame format and 1% used a three-frame format. The sizable minority meant that it was necessary to incorporate multi-frame processing capability into our system development.

5.3.1.2 PICS Usage

We have collected statistical data to gauge the adoption rate of PICS among pornographic Websites. We found that PICS is only adopted by 11% of the pornographic Websites surveyed. Since many of the publishers of pornographic Websites may not want their contents to be filtered out by a Web filtering system, they are reluctant to provide such support in their Websites. We therefore focus on the textual analysis of pornographic Web pages instead.

5.3.1.3 Indicative Key Terms in Textual Context

A Web page that focuses on a major subject carries a specific set of words and phrases that characterize the subject of discussion in the contents provided. This set of terms is usually found common also among other Web pages on the same subject. Therefore, a specific set of terms can be viewed as a unique collection of features characterizing Web pages that emphasize the same subject and lead to a similar theme related to that set of terms. This gives rise to the idea of using a unique set of terms to distinguish a particular type of Web page from others.

This observation is applicable to pornographic Web pages, since these Web pages contain many sexually explicit terms such as "xxx" and "erotic". In order to make use of such sexually explicit terms in the content analysis process of the Web filtering system, it is necessary to compile a list of such terms that appear most frequently among the pornographic Web pages. To avoid introducing too much noise to this list of indicative terms, we need a systematic approach to determine the inclusion of a specific term in the list. We do this by collecting and analysing the statistical data on the usage of indicative terms commonly found in pornographic Web pages.

The indicative terms identified in pornographic Web pages can be classified into two major groups according to their meanings and usage. The group of the majority is comprised of sexually explicit terms which are those with sexual meanings or related to sexual acts, and the other group mostly consists of legal terms which are terms used to establish legitimacy. The reason why legal terms are found on pornographic Web pages is because they tend to have a warning message block in their entry page that states the legal conditions governing the access to the sexually explicit materials contained.

Indicative terms may be found in both the displayed and non-displayed textual contents. The displayed indicative terms may be found in the Web page title, warning message block, graphical text and other viewable textual contents. Non-displayed items are stored in the URL, the meta data of "description" and "keywords" and the image tooltip. We can determine whether some textual contents are displayed or not by checking the markup language tags. For HTML code, displayed textual contents are those not contained within an HTML tag while non-displayed textual contents are found inside an HTML tag. An HTML tag is defined as a block that begins with a '<' and subsequently ends with a '>' in the HTML code of a Web page. For example, the text in Hit control-D to bookmark this site! is displayed, whereas the text in <A HREF="http://www.ntu.edu.sg/"> is not displayed.

5.3.1.4 Statistical Analysis

We have collected a sample of 200 pornographic Websites for statistical analysis based on the discussion of indicative terms given in the above section. A pornographic Website is one whose contents satisfy at least one of the following conditions:

- Sexually oriented contents
- Erotic stories and textual descriptions of sexual acts
- Images of sexual acts, including inanimate objects used in a sexual manner
- Full or partial nudity
- Sexually violent text or graphics

In our study, we are interested in the *number of unique indicative terms* found in the sample Web pages, as well as their *frequency of occurrence* which is given by the number of times a specific term appears in the Web pages. In this research, we focus on English Websites, and English is rich in terms of morphology. So, there are often similar words that can be found among the contents of the sample pages, for example, "pornography" and "pornographic" as well as "sexy" and "sexiest" are not uncommon. Thus, morphed versions of a base word are treated as the same with the base word and contribute to the frequency of occurrence of the base word in the sample pages when collecting the statistical data. Of course, there are a few exceptions. Words such as "sexual" and "sexy", although they are from the base word "sex", are treated as different terms in the data collection process for two reasons:

- The occurrence of the base word is not only common among pornographic Websites, but also among others. One example is "sex" which can also appear very frequently in a health- related Website. Viewing "sexy" and "sexual" differently from "sex" can actually reduce such type of noise incurred by treating all the three terms similarly.
- The morphed version of a base word may have significance as well as a density difference from the base word in different portion of a Web page content. Studying such differences can contribute to more accurate identification of targeted Web pages.

In addition, apart from single-word indicative terms, phrasal terms that comprise two or more words need to be studied. This is because a phrase can give a more specific meaning and more direct indication of the Web page contents than the individual words that make up the phrase. When collecting data of phrasal terms, a phrase containing words similar to another single-word term does not contribute to the statistical data of that term. For example, the phrasal term "adult material" does not contribute to the statistics of the single-word term "adult".

Our study has identified a total of 55 indicative terms comprising 42 sexually explicit terms (e.g. "porn") and 13 legal terms (e.g. "of legal age") with 88.9% and 11.1% frequencies of occurrence. Among the 55 indicative terms, 95.9% are single-word with the remainder being phrasal terms. Table 5.3 summarizes the usage of indicative terms in the eight locations described.

Table 5.3 Usage of indicative terms in eight locations

Location	**Number of unique terms (Total: 55 indicative terms)**		**Frequency of occurrence**
Title of Web page	30	54.55%	6.61%
Warning message block	31	56.36%	10.95%
Other viewable textual contents	45	81.82%	32.19%
Meta data "description"	28	50.91%	7.76%
Meta data "keywords"	39	70.91%	23.66%
URL	30	54.55%	9.91%
Image tooltip	33	60.00%	7.18%
Graphical text	23	41.82%	1.74%

From Table 5.3, more than half of the 55 indicative terms can be found in all of the eight locations except graphical text, with other viewable textual contents and meta data "keywords" containing more than 81% and 70% of the indicative terms, respectively. Further, more than 55% of the indicative term occurrences are found in these two locations. This indicates that the two locations are quite densely populated with indicative terms. The other six locations have their indicative term occurrences ranging from about 1% to 11%. Among these six locations, the graphical text location contributes the least occurrences of indicative terms. Therefore, the probability of graphical text affecting the effectiveness of the textual content analysis capability of a Web filtering system is negligible. Also, only about half of the indicative term occurrences are in the displayed textual contents portion of the sample Web pages, namely, the title of the Web page, a warning message block, and other viewable textual contents. Thus, features in the non-displayed textual contents, which are meta data "description" and "keywords", URLs, and image tooltip, also provide important information on the nature of the Web pages.

5.3.2 System Implementation

We have used the knowledge gained from the analysis of the characteristics of pornographic Web pages to develop an effective Web filtering system. The system architecture is illustrated in Figure 5.1. It consists of two major processes: the Offline Training Process and the Online Classification Process. The Training Process learns from the sample of both pornographic and non-pornographic Web pages in order to form a knowledge base of the NN models. The Classification Process then classifies incoming Web pages from the Web according to the nature of the contents.

The Training Process consists of the following steps: Feature Extraction, Pre-Processing, Transformation, NN Model Generation and Category Assignment. The Classification Process also performs Feature Extraction, Pre-processing and Transformation. In addition, the Categorization step is needed to classify the incoming Web pages based on the results given by the NN models. The Meta Content Checking step performs post-processing to enhance the classification results.

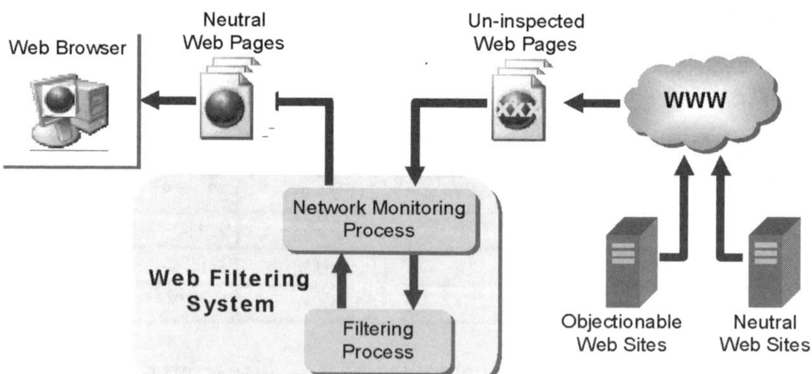

Figure 5.1 System architecture

5.3.2.1 Feature Extraction

A Web page is parsed and the contents in various locations such as the title of the Web page, warning message block, meta data contents of "description" and "keywords", and image tooltips, are extracted as the features to represent the Web page. However, we have decided to exclude URLs due to the difficulties of identifying indicative terms in a URL address.

5.3.2.2 Pre-processing

This step converts all the raw textual contents extracted from the Feature Extraction step into numeric data representing the frequencies of occurrence of indicative terms. It consists of the tokenization of words, and indicative term identification and counting using an indicative term dictionary. Tokenization produces four word lists that correspond to the Web page title, displayed contents, meta contents of "description" and "keywords", and image tooltip. As each list represents a different degree of relatedness to the nature of the Web pages, they will carry different weights when training the NN. As we use frequencies of occurrence of indicative terms in a Web page to judge its relevance to pornography, an Indicative Term Dictionary is employed to support the identification of such terms. The dictionary is compiled according to the results of the statistical analysis. There are two types of indicative terms in the dictionary: sexually explicit terms and legal terms, which collectively give 55 sets of indicative terms. Finally, the Indicative Term Identification and Counting step uses the indicative term dictionary to identify the indicative terms in the four word lists from tokenization, and collects the occurrences for each set of indicative terms in the dictionary.

5.3.2.3 Transformation

The frequencies of occurrence of the respective indicative terms resulted from the Pre-Processing step are then sorted and converted into vectors representing the Web pages which are fed as the inputs to a neural network.

5.3.2.4 Neural Network (NN) Model Generation

Since the engine makes use of NN models (SOM and Fuzzy ART) for the purpose of classification, the networks need to be trained before being used for classification. We have collected a total of 1,009 pornographic and 3,777 non-pornographic Web pages to be used as the training exemplar set for offline NN training. The non-pornographic Web pages are collected from 13 categories of the Yahoo! search engine to cover a wide range of topics. Once the training is complete, the NN generated knowledge is stored in a database along with the dictionary of indicative terms.

5.3.2.5 Category Assignment

The clusters generated from the NN Model Generation are assigned to one of three categories: pornographic, non-pornographic and unascertained, based on a pre-defined assignment strategy. In particular, if a cluster contains at least 80% of Web pages labeled as "pornographic", then the cluster is considered "pornographic". On the other hand, if a cluster contains at least 80% of Web pages labeled as "non-pornographic" then it is considered "non-pornographic". The remaining few clusters are considered unascertained.

5.3.2.6 Categorization

In this step, the incoming Web pages are classified using the trained NN into one of the three pre-defined categories: pornographic, non-pornographic and unascertained.

5.3.2.7 Meta Data Checking

This step checks each Web page classified as unascertained using the contents of the meta data of "description" and "keywords" to determine its nature. The purpose is to further reduce the number of unascertained Web pages. The keywords used are the indicative terms contained in the indicative term dictionary. By analysing and searching for indicative terms within the meta contents, it determines whether a Web page belongs to the pornographic category. If at least one indicative term is found within the meta contents, the associated Web page is classified as pornographic. On the other hand, if no indicative terms are found inside the meta contents, the Web page is identified as non-pornographic Web page. If the meta contents cannot be found or does not exist in the Web page, the Web page will remain as unascertained.

5.3.3 Performance Analysis

First, we measured the performance of the pre-NN steps including Feature Extraction, Pre-Processing and Transformation, which are used for both the Training and Classification Processes. These three steps are responsible for converting a Web page into the corresponding Web page vector. To evaluate their efficiency, we measure the total processing time for the entire training set of 4,786 Web pages. Table 5.4 shows the statistics on efficiency measured for the pre-NN processing.

Table 5.4 Pre-NN processing performance

Measure	Number
Number of Web pages	4,786
Total size of Web pages	93,578,232 bytes
Total processing time	167 s

From Table 5.4, the pre-NN processing steps convert 29 Web pages or 547 KB of data in one second on average. It is important to understand that an efficient pre-NN processing not only helps to improve training efficiency, but also reduces the processing latency in classification.

Also, we observe that each Web page requires an average of 35 ms before reaching the NN and subsequent stages. Using a testing exemplar set comprising 535 pornographic and 523 neutral Web pages, we tested the classification accuracy and efficiency using the two NN models, namely, SOM and Fuzzy ART. Tables 5.5 and 5.6 show the classification accuracy of the SOM and Fuzzy ART models respectively. These tables show that highly accurate classification is possible by combining our knowledge gained in analysing the characteristics of target Web pages (pornographic in this study) with the learning capability of NNs. SOM, in particular, provides an exemplary classification accuracy of 95%, which is much better than the ten popular commercial Web filters that we have surveyed. In addition, the total online processing time is less than 40 ms per Web page on average. This translates to near instantaneous response for highly accurate Web content filtering from the user's viewpoint.

Table 5.5 Classification accuracy of SOM

Web page	**Correctly classified**	**Incorrectly classified**	**Unascertained**	**Total**
Pornographic	508	23	4	535
Non-pornographic	497	7	19	523
Total:	1,005 (95.0%)	30 (2.8%)	23 (2.2%)	1,058

Table 5.6 Classification accuracy of Fuzzy ART

Web page	**Correctly classified**	**Incorrectly classified**	**Unascertained**	**Total**
Pornographic	460	47	28	535
Non-pornographic	483	16	24	523
Total:	943 (89.1%)	63 (6.0%)	52 (4.9%)	1,058

References, Links and Bibliography

[1] http://www.w3.org/PICS/
[2] http://www.rsac.org
[3] http://www.safesurf.com
[4] G.Salton *Automatic Text Processing*. Addison-Wesley, Massachusetts, 1989.
[5] http://www.cyberpatrol.com
[6] http://www.cyber-snoop.com/index.html
[7] http://www.solidoak.com
[8] http://www.symantec.com
[9] http://www.netnanny.com/home/home.asp
[10] http://www.smartfilter.com
[11] http://www.surfwatch.com
[12] http://www.webchaperone.com
[13] http://www.websense.com
[14] http://www.xstop.com
[15] Y. Yang, "Expert network: Effective and efficient learning from human decisions in text categorization and retrieval", *Proceedings of the 17th Annual International ACM SIGIR Conference on Research and Development in Information Retrieval (SIGIR '94)*, Vol. 1, pp. 11–21, 1994
[16] Y. Yang, "An evaluation of statistical approaches to text categorization", *Journal of Information Retrieval*, Vol. 1, No 1/2, pp. 69–90. 1999
[17] Y. Yang, and C.G. Chute, "A linear least squares fit mapping method for information retrieval from natural language texts", *Proceedings of the 14th International Conference on Computational Linguistics (COLING '92)*, Vol. 2, pp. 447–453. 1992
[18] Y. Yang, and C.G. Chute, "An application of least squares fit mapping to text information retrieval", *Proceedings of the 16th Annual International ACM SIGIR Conference on Research and Development in Information Retrieval (SIGIR '93)*, pp. 281–290, 1993
[19] K. Fukunaga, *Introduction to Statistical Pattern Recognition*, 2nd Edition, Academic Press, New York, 1990
[20] G.J. Koehler, and S.S. Erenguc "Minimizing misclassifications in linear discriminant analysis", *Decision Sciences*, Vol. 21, pp. 63–85, 1990
[21] A. McCallum, and K. Nigam, "A comparison of event models for Naïve Bayes text classification", *AAAI/ICML-98 Workshop on Learning for Text Categorization*, Technical Report WS-98-05, AAAI Press, 1998
[22] J. Dalton, and A. Deshmane, "Artificial neural networks", *IEEE Potentials*, Vol. 10, No 2, pp. 33–36, 1991
[23] R.P. Lippmann, "An introduction to computing with neural networks", *IEEE ASSP Magazine April*, pp. 4–22, 1987
[24] P.Y. Lee, S.C. Hui and A.C.M. Fong, "Neural networks for web content filtering", *IEEE Intelligent Systems*, Vol. 17, No. 5, pp. 48–57, 2002.
[25] T. Kohonen, *Self-Organizing Maps*, Springer-Verlag, Berlin, 1995

[26] A. Flexer, "On the use of self-organizing maps for clustering and visualization", *Intelligent Data Analysis*, Vol. 5, No 5, pp. 373–384, 2001
[27] G.A. Carpenter, S. Grossberg, and D.B. Rosen, "Fuzzy ART: Fast stable learning and categorization of analog patterns by an adaptive resonance system", *Neural Networks*, Vol. 4, pp. 759–771, 1991

CHAPTER 6

COMMERCIAL AND INDUSTRIAL APPLICATIONS

6.1 INTRODUCTION

With its emergence as both a popular information repository and communications medium, the Internet has found widespread applications in the industry and commercial world in recent years. The Internet (especially the WWW) offers many advantages for industrial and commercial applications, such as global reach and round-the-clock availability. The Internet is therefore an ideal medium for the dissemination of advertising material (product information, prices, etc.), vendor-buyer communications (e-mail, video conferencing, etc.), payment, delivery tracking, after-sales customer support, and so on. In fact, business-to-consumer (b2c) electronic commerce (e-commerce) has become very popular in recent years. Consumers can now purchase virtually anything online from motor vehicles to foodstuffs, from insurance to magazine subscriptions and from the latest fashion to movie tickets. The range of services and products available is endless.

In addition, multimedia processing, such as image processing, can be combined with Internet technologies to support virtual manufacturing [1]. Multimedia and Internet technologies can be used to provide remote technical support to industrial customers, as well as individual consumers. These technologies can also help in knowledge discovery for decision support, enabling managers to make good business decisions based on past experience.

This chapter uses a number of case studies to illustrate the power of Internet and multimedia technologies for industrial and commercial applications. In particular, we shall study:

- A virtual electronic trading system that supports business-to-business (b2b) e-commerce.
- An intelligent online machine fault diagnosis system that provides after-sales technical support for industrial customers.
- A knowledge discovery system for decision support.

- A Web-based intelligent surveillance system for high-security industrial or military installations.

Although b2c e-commerce has become commonplace in contemporary society, it is business-to-business (b2b) e-commerce that has tremendous potential for growth. An effective virtual electronic trading (VET) framework between business partners can offer a competitive edge against their competitors by supporting just-in-time supply and manufacturing, which minimizes wastage and maximizes efficiency. In Section 6.2, we study a complete business process supported by a virtual environment.

Once sales have been concluded, it is necessary to provide excellent after-sales customer support to build up and maintain strong business relationships. With recent technological advances, it is now possible to provide excellent technical support anywhere, anytime, via the Internet. In partnership with a multinational company, we have developed an effective online fault diagnosis system that helps service engineers either when they offer a help-desk service for their customers, or when they are sent to the customers' sites to identify and rectify machine faults reported by the customers. Section 6.3 describes this system.

At the end of each repair, the service engineer files a service record that includes information such as the name of the company, the name of the service engineer, the machine model, the machine fault and remedial actions, the time since the last service or since installation, the time required to rectify the fault, and so on. All these service records are stored in a database, which can be mined to discover important information that can help managers to make better business decisions. For example, managers can assign tasks to service engineers who are particularly familiar with certain machine models/faults; analysing the faults and frequencies of fault occurrence can lead to better products. We describe a framework for visualizing the multidimensional data for effective decision support in Section 6.4.

Companies and individuals alike take security very seriously these days due to the increase in terrorism and other criminal activities. Many organizations and law enforcement agencies invest heavily in surveillance systems as these systems have been found to be an effective deterrent against criminal threats. However, conventional surveillance systems tend to rely heavily on the constant vigilance of security personnel. Advances in technology have enabled us to develop an intelligent remote security monitoring systems that automates the monitoring task by analysing the monitored scene and generating an alert signal if any suspicious activity is detected. Furthermore, any authorized personnel can perform remote security monitoring anytime anywhere simply by launching a Web browser.

6.2 VIRTUAL ELECTRONIC TRADING FOR B2B E-COMMERCE

Until recently, only large companies could afford to develop and maintain computer systems to automate their trading processes. These *ad hoc* solutions are expensive and preclude widespread applications because they often address specific issues related to the business sectors in which these companies operate. These systems lack flexibility (e.g. for diversification of business activities) and scalability (e.g. for subsequent expansion), making them difficult to maintain.

With the declining costs of computers and the emergence of the Internet, it is now possible to develop a generic electronic infrastructure for virtual trading. We have conducted a survey of existing b2b electronic trading systems so as to identify the key attributes of these systems, which highlight specific areas for further improvements. These findings form the basis of our VET system, which provides a generic framework for a variety of business models and practices.

The VET system provides an electronic process for purchasers and suppliers to advertise or search for specific products and execute an order and payment. It also allows the purchasers to monitor the status of active processes. The VET system architecture includes different components, such as Purchaser Interface, Supplier Interface, Search Engine, Order Management and Payment Engine for completing the b2b procurement process. In doing so, VET system users are able to fulfil their orders by using a Web browser over the Internet.

6.2.1 Survey of b2b E-commerce Systems

Several factors are chosen to assess the value and efficiency of different b2b e-commerce systems. These factors include service topology, products/services provided, payment methodology, search engine and other features.

- *Service Topology* – What kind of services does a b2b trading system provide? One-to-many (1-to-N) or many-to-many (N-to-N)? 1-to-N means that one supplier provides b2b e-commence services on its site for its sub-suppliers or retailers. On the other hand, N-to-N topology refers to the services that a third party provides for many suppliers and many related purchasers. Therefore, N-to-N allows purchasers to easily compare similar products among different suppliers.
- *Products / Services Provided* – Some systems, especially 1-to-N systems, only supply the products that they produce. Some N-to-N systems deal with specialized products such as chemical reagents or electronic components offered by multiple suppliers. Other N-to-N systems aim to address the needs of such products as maintenance, repair and operational (MRO) items that make up 80% of the trade [2].

- *Payment Methodology* – Open accounts are typically used for purchasers with established credit and are the standard business practice between trading partners. Normally, the payment is made within a predefined time period from the date of shipment. On the other hand, credit card payments are also common among companies without established credit arrangements.
- *Product Search* – This refers to the search facility for finding specific products easily and quickly. A clear and efficient search for finding specific products not only promotes a satisfactory user experience; it also delivers value to purchasers in the form of time saving.
- *Other Features* – These include information updating, system integration and security. Some existing b2b e-commerce systems send newsletters and e-mailse-mail to inform interested clients once new information is available. System integration [3] allows b2b e-commerce systems to integrate with available commercial systems; databases or custom-designed solutions already used by different suppliers or clients. A logical extension of this is the streamlining of business practices, for example, approval routing, tender process, and so on. Security is a major issue that b2b e-commerce systems have to address.

Table 6.1 shows a comparison of representative b2b e-commerce systems based on the features that they possess. With regard to the Service Topology, both 1-to-N and N-to-N types are well represented. 1-to-N systems are typically managed by well-known manufacturers such as Cisco Systems [4], Dell Computer Corp [5] and GTE Corp [6]. These multinational companies make use of 1-to-N b2b service to promote their own products in a cost-effective way. In each case, however, only one supplier provides the products so that it is difficult for clients to compare products from different suppliers.

Table 6.1 Survey of b2b E-commerce systems

System	Service type	Product	Payment	Product lookup	Other features
Cisco Systems	1-to-N	Cisco products	Credit card	Search engine catalogue browsing	Newsletter integration with Oracle database
Dell Computer Corp.	1-to-N	Dell computer products	Credit card	Search engine	SSL monthly newsletter
GTE Corp.	1-to-N	Telecom services	Open account	Catalogue browsing	SSLand 128-bit RC4 encryption

System	Service type	Product	Payment	Product lookup	Other features
Chemdex Corp	N-to-N	Chemicals	Open account credit card	Search engine	Integration with ERP and SAP Netscape's Secure Commerce ServerTechnology
Trade-Out.com	N-to-N	Business surplus and features products	According to different suppliers	Search engine catalogue browsing	Newsletters
E-steel Corp.	N-to-N	Steel	Open account	Search engine	Newsletters

In contrast, companies that adopt the N-to-N Service Topology may sell a variety of products such as TradeOut.com [7], or they may specialize in certain kinds of products. For example, Chemdex Corp [8] specializes in chemical reagents, whereas E-Steel [9] specializes in steel products. These companies supply to niche markets such as research organizations as in the case of Chemdex Corp. The main advantage of using N-to-N services is that clients could compare the products offered by different suppliers on the same site through the market-maker before making a purchase. However, relatively few b2b systems offer general products. This suggests that the advantage of adopting the N-to-N market-maker topology has not been fully realized.

The common methods of payment are open account and credit card. All the systems surveyed provide a search engine to facilitate product search. In addition, a few companies also provide catalogue-browsing facilities to complement their search engines. These are currently two of the most popular methods used for specifying the required product. In addition, a good design of Web pages can make the search more efficient.

Many companies maintain databases to keep information on their trading partners, as well as on their own products. The most common modes of advertising are advertisements on their own Websites and regular newsletters sent via e-mail to current and potential clients. Moreover, some of these systems (i.e. Dell Computer Corp. and TradeOut.com) allow clients to customize their newsletter. To make the system friendlier to users, Chemdex Corp. allows third party systems to be integrated into its b2b service.

All of the companies that were surveyed regard security as a major issue in their system design. Since the Internet cannot guarantee secure data transmission, additional security technologies need to be adopted. The secure socket layer (SSL) protocol is commonly used, though GTE, for example, also employs the 128-bit RC4 encryption technique to meet the security requirements of the US federal government.

Up to now, many existing b2b e-commence systems have been implemented for selling and buying products online. However, these systems have advantages and disadvantages as discussed above. Moreover, few documents are available on the design details of system architecture. Section 6.2.2 describes a VET system that has been developed to overcome the limitations of existing systems by incorporating advanced features into the system.

6.2.2 The VET System

There are two main types of b2b trading activities. The first type involves specialized items such as electronic components or chemical reagents as required by specialists in, for instance, a research-oriented establishment. This mode of trading is characterized by the requirement of expert knowledge necessary in deciding what products to purchase. Often, there are a few companies that supply these kinds of products. Using expert manpower to undertake such mundane activity as product procurement in the traditional way is a costly exercise, so automation would greatly benefit organizations that are engaged in this mode of trading.

The second mode of trading between businesses is commonly referred to as the MRO (maintenance, repair and operation) type, where low-unit-cost, high-volume and non-mission-critical products are traded. This type of trading is said to account for 80% of the total b2b transactions [2]. Consequently, much work has focused on this mode of b2b trading [10]. However, a comprehensive solution has yet to emerge.

This is where VET comes in. It is developed to provide a complete solution to the problems that exist today in the procurement process. The generic and cost-saving nature of the VET system should give it a wide appeal among business users. Figure 6.1 shows the VET system architecture. The major system components are described in Section 6.2.3 VET provides Purchaser and Supplier Interfaces for participating enterprises. In some cases, a company may act as both the supplier and the purchaser of goods within the supply chain. VET distinguishes the two only on an in-need basis, so the role of a company may change accordingly.

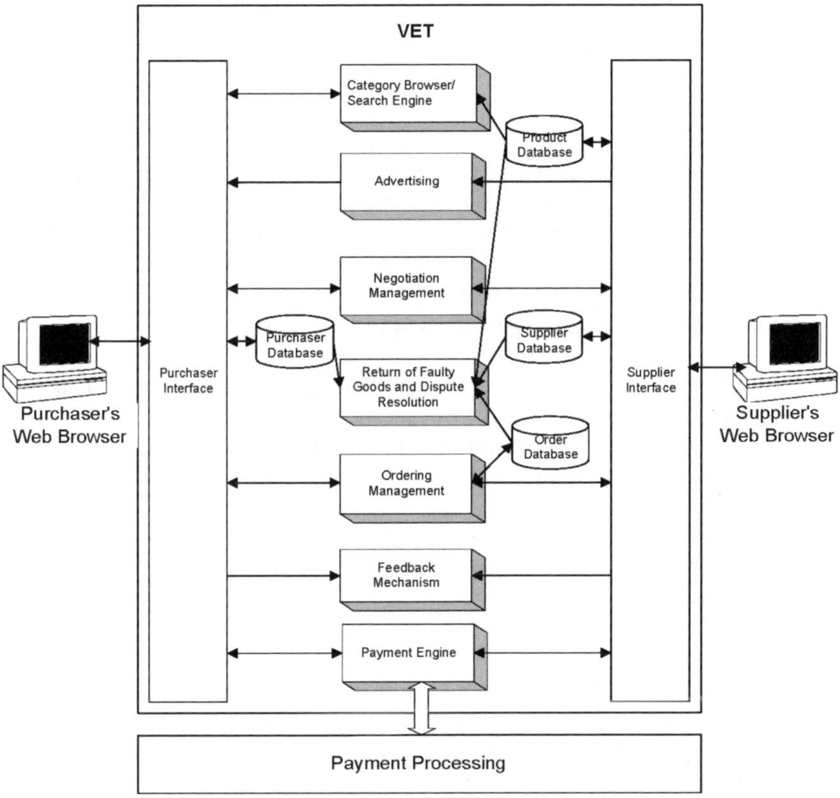

Figure 6.1 VET system architecture

6.2.3 VET System Components

6.2.3.1 User Interfaces

Two user interfaces are provided: one for suppliers and one for purchasers. The interfaces are designed as the upper application layers for the VET, so they hide the complexity of the lower layers from users who can execute their procurement processes simply. Both interfaces are password protected so that users need to logon to the system before use.

The interface for purchasers mainly includes product-searching pages, ordering pages and payment pages. These pages are designed to link purchasers with a well-structured Website managed by the central VET process.

The product-searching pages include a search criteria page and a search result page. The search criteria page contains options such as keyword, supplier name, product name and part number to help purchasers find the products they are interested in. The results are displayed on the search result page. This page lists all the matching products for the purchaser.

The ordering pages include three Web pages: order placing page, order confirmation page and order status page. Order placing page enables purchasers to place their orders via a Web browser. Purchasers can change and confirm the orders that they have placed on the order confirmation page. The order status page allows purchasers to check the current status of their orders in real time. Payment pages are used for purchasers to pay for the ordered products online. The purchaser is required to enter credit card or open account information as appropriate.

The interface for suppliers includes product service pages and an order-listing page. Product service pages allow suppliers to update and change the information on their products. Confirmation is required each time a supplier wishes to change or update the information. The order-listing page is used for displaying all the information regarding current orders and order history. It is very useful for suppliers to know about the situation of how their products are selling as soon as possible.

6.2.3.2 Advertising

Advertising is the start of many business transactions. Suppliers need to reach out to the business community to advertise their products and services. Envisaged as a market-maker, the VET system allows suppliers to advertise their products and services to a wide range of trading enterprises economically via the Internet. Suppliers and potential purchasers form dynamic supply chains through the VET system, realizing the N-to-N topology.

Selling to business customers requires an enterprise-wide solution. It starts with customized catalogues for different major trading partners. The catalogues must have detailed, up-to-date and accurate information organized in such a way that the search for products/services will be fast and easy (see Section 6.2.3.3). This is particularly important for specialized products such as chemical reagents and electronic components because expert manpower is costly for carrying out the tasks of finding the appropriate items to purchase. In the case of MRO type of products, there are even more opportunities for cost savings through automation and pre-negotiations (see Section 6.2.3.4).

6.2.3.3 Catalogue Browser/Search Engine

Figure 6.2 illustrates the Search Engine environment that allows users to search for specific products. As shown, a purchaser enters information such as keyword, product name or CAT/part number and supplier name using a Web browser through the Internet. On receiving the request, the Search Engine of VET executes a search according to the criteria that the purchaser entered. The Search Engine generates a query to retrieve information from a permanent database called *Product Database*. If there is a match, the Search Engine will return the results to the purchaser. Otherwise, the purchaser will be prompted to search again using a set of new criteria. The search engine also stores the most recent or most frequent hits to expedite searches on a statistical basis.

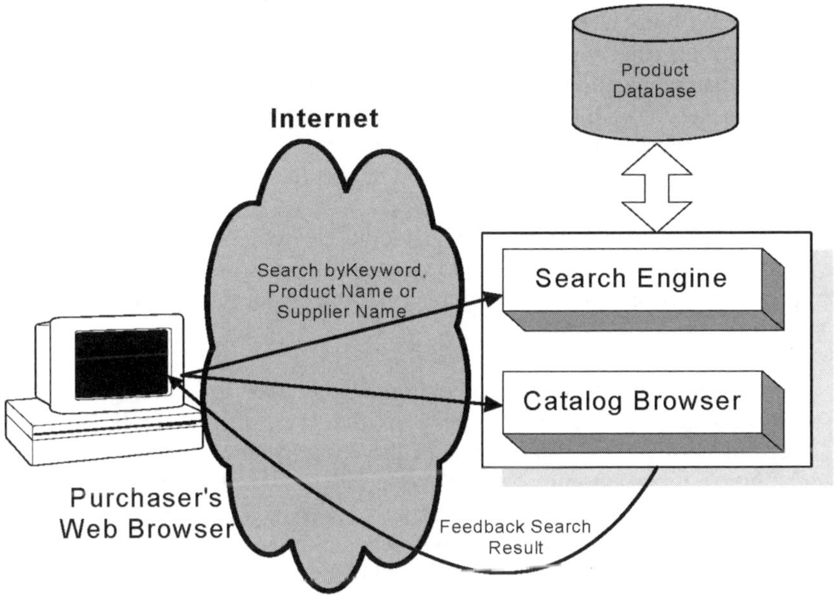

Figure 6.2 Catalogue browser/search engine environment

Apart from the query-based search engine described above, the VET system provides an alternative method of finding suppliers and/or products/services. This is a catalogue browser with similar products grouped together under such headings as "electronic components" or "computer accessories". The browser is organized in the form of a tree-structure for navigation and has cross-links. The concept is similar to the directed acyclic graph (DAG) described in Ref. [11]. The VET system stores a history of paths visited and provides a quick way of navigating between these paths.

6.2.3.4 Negotiation Management

Customer loyalty is often much valued (e.g. airlines and petrol companies run frequent customer programs), so suppliers tend to offer better deals to purchasers with guaranteed minimum quantities. The market-maker nature of the VET system allows companies to pre-negotiate deals on an in-need basis. Some companies may also form strategic alliances to benefit from the economy of scale and to ward off any potential hostile competition. Having pre-negotiated deals should enable each supplier to have a better idea of how much stock to keep and when it might be needed. All this would minimize wastage and the savings could be passed on to the purchaser (as an incentive to remain business partners).

In addition to the type of pre-negotiation described above, the VET system also supports different modes of negotiation, such as bidding and bargaining. In the former case, competing suppliers or purchasers may bid for the best deal for them where multiple companies are usually involved at the same time. In the latter case, bargaining usually takes place between a pair of suppliers/purchasers at a time. In any case, the system employs the same Negotiation Management technology for negotiation after an active search conducted by a purchaser.

If a purchaser receives a quotation and wishes to negotiate certain aspects of it, a more acceptable alternative is formulated. For instance, the purchaser may wish to get a lower price or a better delivery arrangement through negotiation with the supplier.

The negotiation management environment is illustrated in Figure 6.3. The major advantage of this environment is the use of a Web-based Video-Conferencing System [12]. This is one of the few aspects of the VET system that requires human presence in real time. The aim is to provide a virtual face-to-face negotiation environment conveniently at an optimal cost versus quality compromise. In VET, the Web-based video-conferencing system is used for negotiations only if a virtual face-to-face negotiation is considered necessary. Once properly set up, other processes are automated without the need for human intervention.

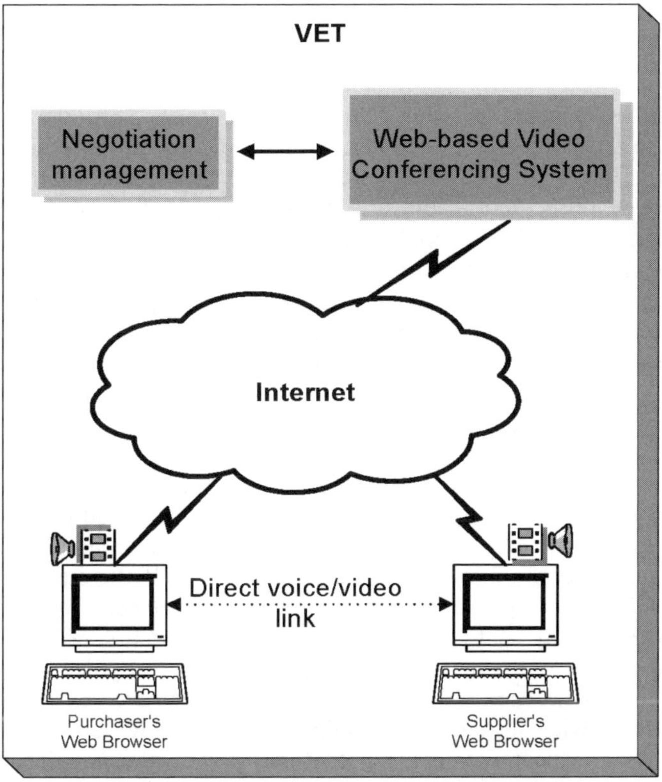

Figure 6.3 Negotiation management environment

6.2.3.5 Ordering Management

Ordering Management comprises four components: Order Initialization, Order Verification, Order Acceptance and Order Monitoring. An Order Database is integrated into the Ordering Management to store information related to each order.

- *Order Initialization* – An order request may be automatically initiated by companies having an automatic inventory control system. When stocks run low (< x% as determined by the company), an order request may be initiated and sent to the VET system. If there are established links to suppliers, the VET system will pass the request to them. Otherwise, VET will manage the search and/or negotiation processes. The order record is stored in the order database.
- *Order Verification* – The VET system then checks all the fields such as the expiration date, goods description, terms and conditions and the price

in the order record given by purchaser. If the order is not accurate, the system will ask the purchaser to refill the order form.

- *Order Acceptance* – If the order passes Order Verification, the VET system will accept this order. The system will then inform the purchaser and supplier to complete the order. At any time, either party might decide to abort the order. The VET system has exception handling capabilities that include proper termination and garbage collection.
- Order Monitoring – This is an important aspect of the whole ordering procedure and may offer some legal protection (in the form of evidence of transactions) for legitimate parties in case of dispute. It allows VET users to monitor the status of their orders in real time through the Internet. Order Monitoring reads and writes data from/to the Order Database.

6.2.3.6 Payment Engine

The Payment Engine is closely related to product delivery. Depending on the actual agreement between companies, products may be delivered before, after or at the same time as payment. The payment engine receives reports containing the status of each order from the Ordering Management. If the status is complete, which is a successful sign verified by the Order Acceptance, the central process would activate the Payment Engine. The Payment Engine first provides a form for the purchaser to select the method of payment (if not already decided prior to this step). Two methods, Credit Card and Open Account, are allowed by the VET system. Purchasers with special relationships established by suppliers are able to complete their payment within the predefined period after shipment. After the payment is verified, the supplier arranges shipment accordingly.

6.2.3.7 After-Sale Service and Dispute Resolution

The VET system can act as a mediator in the event that a dispute arises after transactions have been made through the system. In most cases, however, clear-cut issues such as the return of faulty goods can be done through the VET system's negotiation management environment, especially when most agreements should already contain clauses regarding such issues. In rare cases, legal and/or government bodies can be involved by activating VET system's dispute resolution feature. Evidence, in the form of time-stamped records of transactions, agreement documents, and so on, are also important. Obviously, the dispute resolution module will have read-only access to the appropriate databases.

6.2.3.8 Security

Security is a major issue in any system that handles sensitive information. The VET system safeguards the integrity of sensitive data such as trade secrets and other propriety information so that only the parties concerned can access the information. This is done by using the latest techniques available, such as the Secure Socket Layer (SSL) protocol for the transmission of sensitive information and access to sensitive information is through proper authentication procedures.

6.2.3.9 Discussion

The VET system has been developed as a centralized market N-to-N VET framework that facilitates efficient and economical b2b trading, as well as addressing some of the issues faced by other similar systems. The system uses a main process to manage and schedule sub-processes to complete the various tasks needed in the procurement process through its purchasers' and suppliers' interfaces. It allows users to conduct efficient and secure transactions without having to worry about the internal workflow of the various sub-processes. The VET system also encourages cost-saving business practice, such as bulk purchasing and the formation of strategic alliances.

The flexible nature of the VET system means that it could be used both for specialized products as well as for MRO items. Another advantage of adopting a flexible generic framework is that it does not preclude the introduction of more advanced technologies when these become available. For example, intelligent mobile agents could later be employed to take a market-maker role within the VET paradigm. The various components that make up the VET system demonstrate how the Internet can be effectively used both for information dissemination (e.g. advertising) and communications (e.g. negotiation) for commercial applications.

6.3 Web-Based Customer Technical Support

Customer service support is becoming an integral part of many multinational manufacturing companies. These companies generally have a customer service department that provides installation, inspection and maintenance support for their customers. A help-desk service centre is usually established to answer frequently encountered problems from the customers. Service engineers from the help-desk centre respond to customers' enquiries via telephone calls and carry out on-site repair if necessary. At the end of each service, a customer service report is generated to record the problem and the remedies taken to rectify the problem. These reports are then stored as service records in a customer service database.

Traditionally, the identification of machine faults relies heavily on the expertise of the service support engineers. It is often a burden on the company to train and retain a large pool of expert service engineers. Since the customer service database serves as a repository of invaluable information that can be used for machine fault diagnosis, the customer service database can be mined to support customer service activities over the WWW.

6.3.1 Customer Service Database

Each time a service engineer completes a repair, a service record is filed to document the fault(s) and remedial actions taken to rectify the fault(s). Service records are stored in the customer service database to keep track of all reported problems and remedial actions. Each service record consists of customer account information and service details: *fault-condition* and *checkpoint information*. Fault-condition contains the service engineer's description of a machine fault. Checkpoint information indicates actions taken to rectify the fault. Service records are therefore useful for future fault diagnosis.

Fault-condition	3008 PCB CARRY MISS ERROR. PCB WAS NOT TRANSFERRED BY THE CARRIER DURING LOADING BUT STAYED AT THE DETECTION POSITION OF PCB DETECTION SENSOR 2.	
Checkpoint group: AVF_CHK007		
Priority	Checkpoint description	Help file
1	CONFIRM WHETHER THE CARRY GUIDE PINS ARE IN LINE WITH PCB.	AVF_CHK007-1.GIF
2	CONFIRM WHETHER THE PCB IS IN CORRECT DIRECTION.	AVF_CHK007-2.GIF
3	CONFIRM THE POSITION OF THE GUIDE LOWER LIMIT SENSOR. (I/O 0165)	AVF_CHK007-3.GIF
4	CONFIRM THE TIMING FOR PCB 2 DETECT SENSOR.	AVF_CHK007-4.GIF

Figure 6.4 Sample service record

Figure 6.4 shows an example of a fault-condition and its checkpoint information found in a service record. It contains a checkpoint group name, and prioritized list of checkpoint descriptions. Each fault-condition as a unique number associated with it so that it can be used as a key in the customer service database. The checkpoint solutions are prioritized to guide the service engineer through the possible solutions from the most probable to the least probable. In addition, each checkpoint description has an associated help file that provides graphical information to help the service engineer to locate the fault and carry out the suggested remedial action. An example help file for the second checkpoint description in Figure 6.4, which is stored as AVF_CHK007-2.GIF, is illustrated in Figure 6.5.

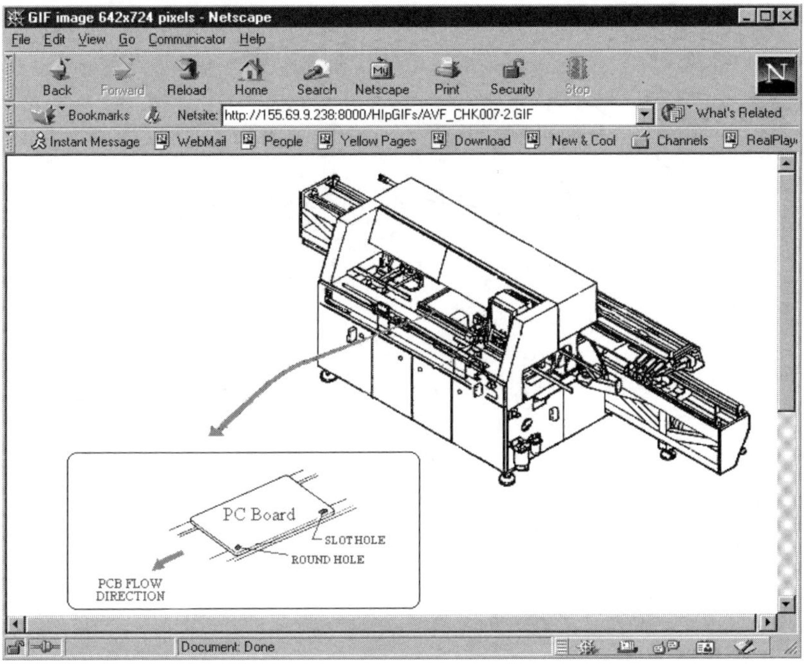

Figure 6.5 Help information to assist user carry out the remedial task

Data are stored as unstructured text in the machine fault and checkpoint tables. There are over 70,000 service records in the customer service database with over 50,000 checkpoints. In addition, structured data on over 4,000 employees, 500 customers, 300 different machine models and 10,000 sales transactions are also stored. A new technique has been developed specifically for mining the unstructured fault-conditions and checkpoints data for machine fault diagnosis.

6.3.2 Data Mining for Machine Fault Diagnosis

Figure 6.6 shows a framework of the intelligent data mining process comprising the offline Knowledge Extraction Process and the online Fault Diagnosis Process. The Knowledge Extraction Process is expanded in Figure 6.7, which shows how useful information is extracted from the customer service database to form a knowledge base that contains neural network models and a rule-base.

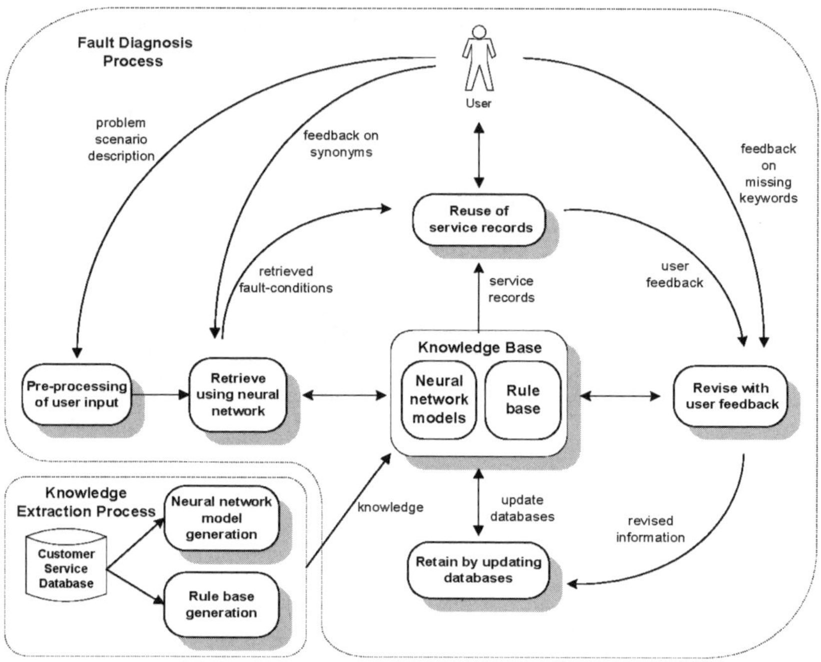

Figure 6.6 Data mining process for machine fault diagnosis

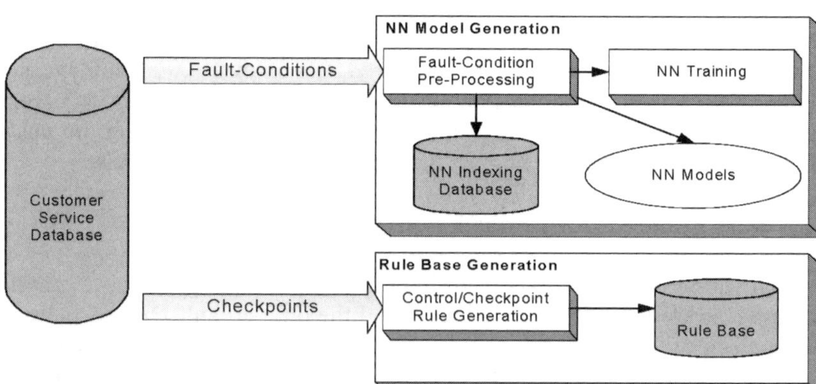

Figure 6.7 Knowledge extraction

In our implementation, we have compared the performance of two neural network models: supervised learning vector quantization (LVQ) and the unsupervised Kohonen's self-organizing map (SOM) [13]. The rule-based inference engine has been developed under the C Language Integrated Production System (CLIPS) environment [14], which provides a step-by-step guide to help the user identify the fault(s) in question.

The neural network models and the rule-based work within the case-based reasoning (CBR) [15] cycle to support online Fault Diagnosis Process. The Fault Diagnosis Process uses the four stages of CBR cycle: retrieve, reuse, revise and retain to diagnose customer reported problems. It accepts the user's problem description as input, maps the description into the closest fault-conditions of the previously reported faults from the knowledge base, and retrieves the corresponding checkpoint solutions for the user to try to resolve the current problem. The user's feedback on the Fault Diagnosis Process is used to revise the problem and its solution. The revised information is retained as knowledge for enhancing its performance in future diagnosis.

6.3.3 Machine Fault Diagnosis over the WWW

Figure 6.8 shows the Web-based machine fault diagnosis system that runs in the Windows NT 4.0 environment [16].

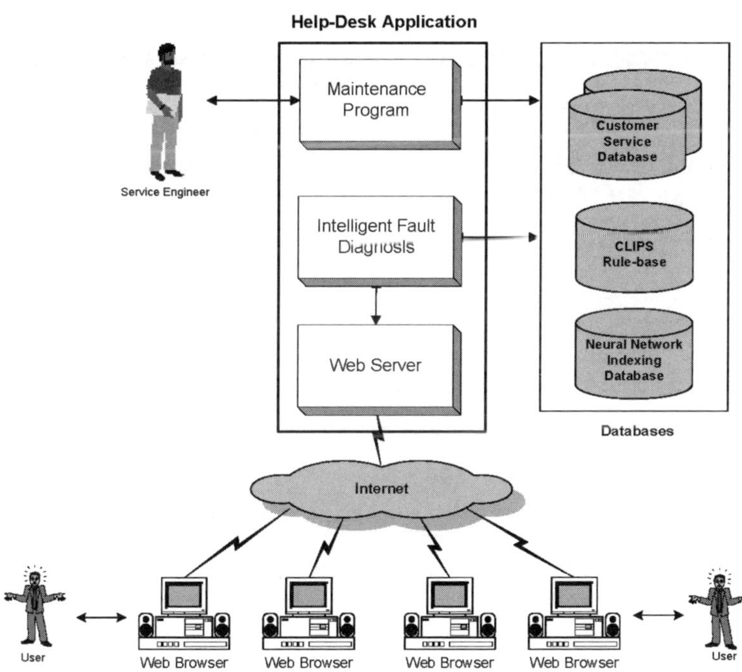

Figure 6.8 Web-based system for machine fault diagnosis

The data mining technique described in Section 6.3.2 is implemented to perform fault diagnosis. The Netscape Enterprise Server is used as the Web Server. Service engineers can interact with the Maintenance Program for updating service records. The Microsoft Access database management system is used to implement the Customer Service Database. The rule-base is written in CLIPS format. Hypertext Mark-up Language (HTML) is used to create the user interface as Web pages to accept the user's fault description. Customers can connect to the Web Server using any Web Browser.

Figure 6.9 shows the Web-based interface for accepting user input of a fault description. If the error code is known, no other information is then required from the user for further processing. The corresponding fault-condition can be identified and its checkpoints can be retrieved. Otherwise, the fault description can be entered in natural language or as a set of keywords. The user can also provide the names of machine components and their states as input as shown in Figure 6.10. If the user input contains keywords that are not in the predefined keyword list, synonyms of these keywords will be retrieved for user confirmation as input keywords. We have adopted Wordnet's dictionary [17] for textual information processing, such as finding the synonyms of keywords.

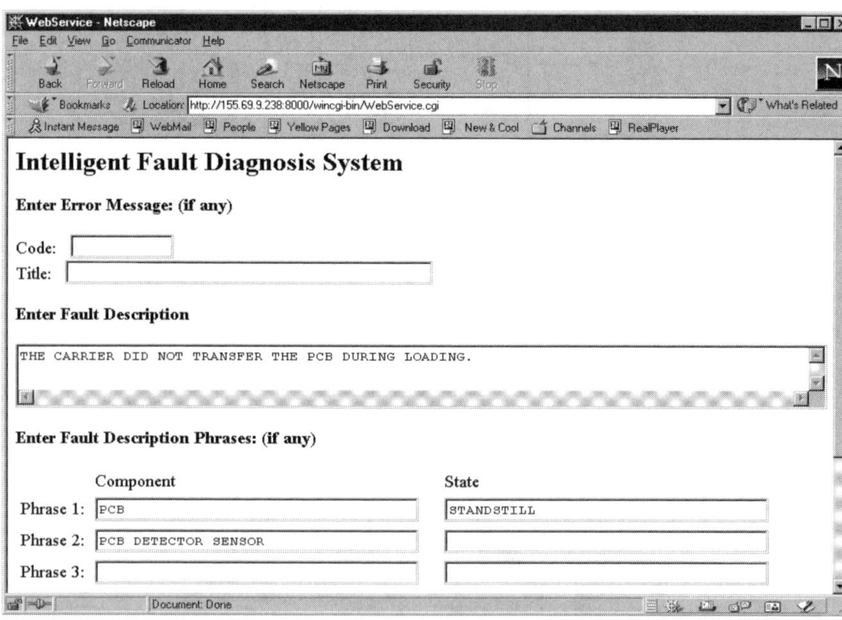

Figure 6.9 User interface for entry of fault descriptions.

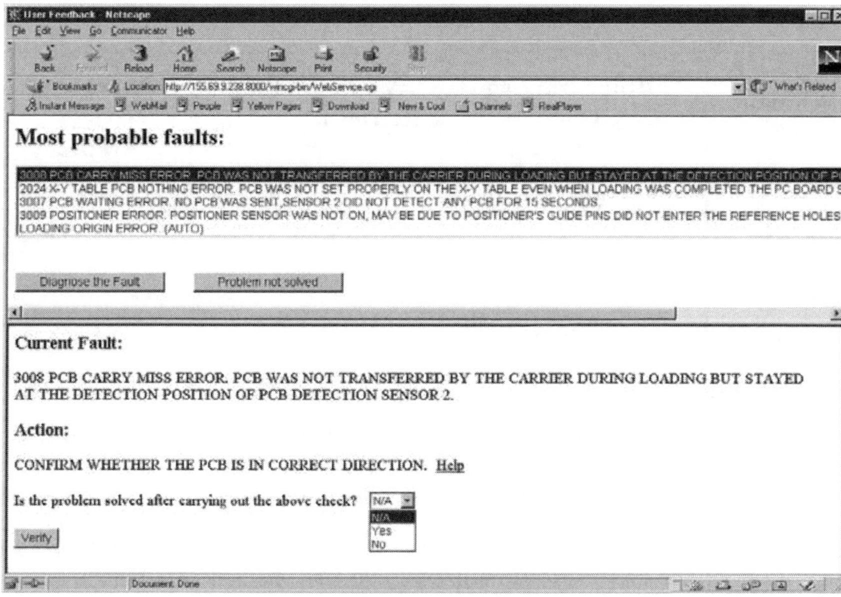

Figure 6.10 Retrieval results in response to user input.

6.3.4 Performance Evaluation

The system described in Section 6.3.3 has been field tested by service engineers of a multinational company that manufactures, supplies and services a range of machin ery for the electronics industry. The two neural networks, LVQ and SOM, are also compared with the k Nearest Neighbour (kNN) technique used in traditional CBR systems for retrievalas seen in Table 6.2. A traditional CBR system using the kNN technique needs to store all the cases in the case-base in order to perform accurate retrieval. The use of neural networks with CBR greatly reduces the search space due to generalization of knowledge through training.

Table 6.2 Comparison of different retrieval techniques

Retrieval Technique	Offline training Time	Average online Retrieval Time (s)	Average retrieval Accuracy
kNN	–	15.3	81.4%
LVQ	96m44s	1.9	93.2%
SOM	264m35s	0.8	90.3%

kNN stores cases in a flat memory structure, extracts keywords from the textual descriptions and uses normalized Euclidean distance for matching. It always assigns equal weights to the individual attributes (i.e. keywords). Therefore, retrieval is less accurate. Moreover, the major drawback of the kNN technique is that new cases retained are indexed separately into the flat memory structure. This causes the search space to keep increasing, further reducing the efficiency.

Although the total training time for each neural network is quite high, it is still acceptable, as the training is carried out offline only once. In addition, the average online retrieval time for each neural network is quite fast. The SOM neural network requires a longer training time, but, it performs more efficiently online compared to the LVQ neural network. Overall, service engineers have found the Web-based fault diagnosis system to be easy to use, which offers fast and accurate retrieval.

6.4 Knowledge Discovery for Managerial Decisions

The customer service database described in Section 6.3.1 not only provides useful information for service engineers in fault diagnosis, but also contains valuable information that could be used for future decision-making. For example, analysis of the service records may reveal patterns that suggest certain design faults inherent in some products; perhaps some service engineers are better at handling certain faults or certain products.

However, finding the relevant information (identifying trends, patterns, etc.) is difficult because corporate databases, such as the one described in Section 6.3.1, tend to be very large. In addition, the data stored is often multidimensional, and can be incomplete, making knowledge discovery even more difficult. So, how do we discover useful information hidden in a large corporate database?

Online analytical processing (OLAP) is an efficient way to access a large database for multidimensional analysis and decision support, but OLAP techniques alone cannot derive patterns from the stored data – a capability most companies need if they are to identify useful trends and patterns.

Data mining, in which analysts build a data cube to represent the data at different levels of abstraction, is therefore a natural partner to OLAP. Analysts can then use OLAP techniques to visualize the data cubes and identify interesting patterns, trends, and relationships among them. Therefore, data cubes make it easier to use OLAP for large multidimensional databases, and OLAP makes it easier to analyse the data cubes themselves.

To capture this partnership, we have designed a seven-step data mining process for decision support that combines OLAP and data cube analysis with major data mining functions. As a test of its effectiveness, we applied it to the customer service department described in Section 6.3 Feedback from service engineers and customer service administrators has been encouraging. The process is flexible enough to suit any company that needs to analyse a large multidimensional database, and it works with most popular data mining tools.

6.4.1 Seven-Step Process for Knowledge Discovery

Our process for knowledge discovery has been developed in collaboration with a multinational company that manufactures, supplies and services insertion and surface-mount machines primarily for the electronics industry. The seven steps in our process are as follows [18]:

- *Establish mining goals* – Consider the cost-benefit tradeoff and the expected accuracy and usefulness of the results.
- *Select data.* – Decide which kinds of data are useful, which are worth considering and how big the sample size should be.
- *Preprocess data.* – Filter out noisy, erroneous or irrelevant data and develop strategies to handle missing data.
- *Transform data.* – Where appropriate, reduce the number of data attributes or introduce new ones based on existing data attributes. Combine data tables and project the data onto working spaces, that is, tables that represent the optimal abstraction level for the problem of interest.
- *Store data.* – Integrate and store data at a single site under a unified scheme.
- *Mine data.* – Perform appropriate data mining functions and algorithms according to the mining goals established in step 1. Typically, analysts first

construct data cubes to provide multidimensional views of the data. They then perform online analytical mining using the multidimensional data cube structure for knowledge discovery.

- *Evaluate mining results* – Perform various operations such as knowledge filtering from the output, evaluate the usefulness of extracted knowledge, and present the results to the end-user for feedback. The results in this step may lead to changes in earlier steps.

6.4.2 Establish Mining Goals

Our partner company has identified six major mining goals, categorized according to the company's four business areas (marketing, manufacturing, resource management and customer support). Since these areas are interdependent, the data and mining goals can overlap.

1. Identify poor selling machine models and the underlying reasons for poor sales. (Marketing)
2. Identify suitable customers for cross sales. (Marketing)
3. Identify the machine models that exhibit the most faults. (Manufacturing)
4. Identify good service engineers. (Resource Management)
5. Identify the best matches between engineering expertise and machine models on the basis of how individual engineers have performed to date. (Resource Management)
6. Identify the customers who have reported the most faults recently and therefore possibly require more attention. (Customer Support)

6.4.3 Select Data

Currently, the customer service database contains over 70,000 service records and information on more than 4,000 employees, 500 customers, 300 different machine models and 10,000 sales transactions. Figure6.11 shows the six tables in the database. The MACHINE_FAULT and CHECKPOINT tables contain unstructured textual data on machine faults and the corresponding checkpoints/remedies. These tables are for fault diagnosis described in Section 6.3, not for decision support. The other four tables store structured data information on customers (CUSTOMER), employees (EMPLOYEE), sales (MACHINE), and maintenance (SERVICE_REPORT). Each table has several attributes, but not all are suitable for mining.

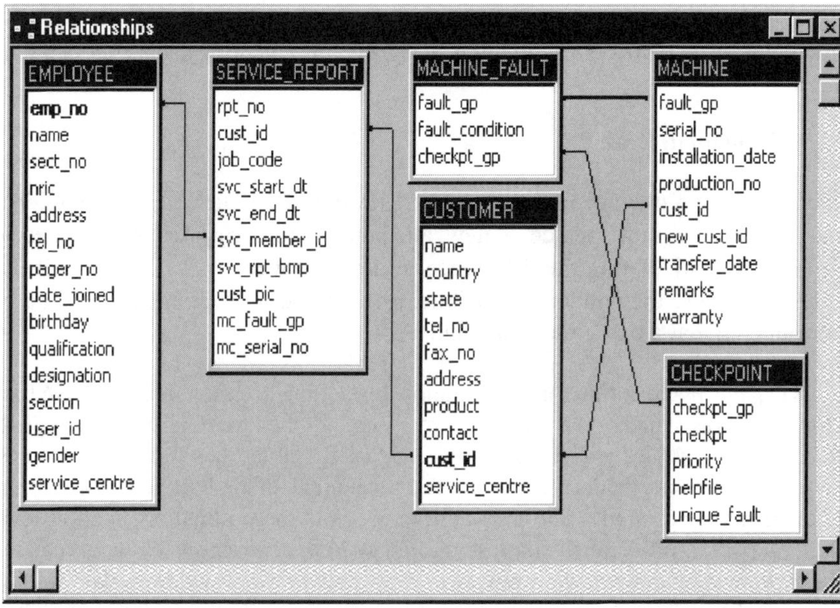

Figure 6.11 Relationships between tables in the customer service database

The EMPLOYEE and CUSTOMER tables are not suitable for mining because they contain categorical attributes. These are attributes such as employee number, customer telephone number, customer identification number, and so on, all of which are characterized by having distinct, and often fixed values. Clearly, using categorical attributes for mining is unlikely to yield useful information.

Therefore, only two of the six tables are actually used for mining useful information. MACHINE, which contains about 10,000 records of machine sales could be used to address mining goals 1, 3 and 6. For example, attributes such as `fault_gp` (fault group) and `installation_date` are likely to be useful. Similarly, the SERVICE_REPORT table contains useful attributes such as `mc_fault_gp` (machine fault group), `svc_start_dt` (service start date) and `svc_end_dt` (service end date), which could be used to address mining goals 1, 2, 4, 5 and 6.

6.4.4 Preprocess Data

Data suitable for mining typically come from different sources, have different formats, and can have missing or incorrect values. Incorrect data entries, though relatively rare, can distort final results. Mining tools support preprocessing to some degree (mostly in the form of noise filtering), but an analyst must interpret and reject other erroneous inputs. Some simple rule-based methods can partially automate problems that are easy to detect, such as improper dates for example, 30th February. Preprocessing also involves eliminating categories in categorical data to facilitate visualization.

6.4.5 Transform Data

We have made two transformations to the SERVICE_REPORT table to make data mining easier. We first created a new attribute, `svc_repair_time` (service repair time), by calculating the difference in days, between `svc_start_dt` and `svc_end_dt`. We then added an attribute `months` to indicate the months before the current date that a machine fault occurred.

We have made a third transformation by adding a new table, MC_SALES (machine sales), which was created by joining the MACHINE and SERVICE_REPORT tables. The new MC_SALES table has three attributes: `mc_model` (machine model), `num_faults` (number of faults), and `num_sold` (number of machines sold). The new table represents the optimal abstraction level for the company's problem of interest that is, to look at machine faults associated with a particular model and the number of that model sold (to address data mining goals 1, 3 and 6). The original MACHINE and SERVICE_REPORT tables and their attributes remain intact.

6.4.6 Store Data

We use a method known as data warehousing to store the data for subsequent mining. The purpose of a data warehouse is to make standard OLAP operations and data mining easier. We have stored the data collected from the various tables into a data warehouse, from which OLAP data marts are generated. OLAP data marts are subdivisions of data that capture the abstraction level of interest. The three data marts are: SERVICE_REPORT, MACHINE, and MC_SALES, which correspond to the useful tables described above. Data cubes are then constructed from these data marts to obtain multidimensional views of the data.

6.4.7 Mine Data

Data mining is performed in two phases: the construction of data cubes and the application of mining functions to those cubes. Data cube construction begins with the selection of the appropriate attributes from the data marts. The data cubes present information in three-dimensional space at various conceptual levels. Table 6.3 shows the data marts we have used to generate the data cubes and the associated attributes.

Table 6.3 Data marts and attributes used to construct data cubes

Data mart	Attributes	Data cube
SERVICE_REPORT	mc_fault_gp svc_start_dt svc_repair_time svc_member_id cust_id months	SVC_RPT_CUBE
MACHINE	fault_gp instal- lation_datecust_i d	MACHINE_CUBE
MC_SALES	mc_model num_faults num_sold	MC_SALES_CUBE

Figure 6.12 shows a three-dimensional view of the SVC_RPT_CUBE where the three dimensions mc_fault_gp (machine fault group), months and svc_member_id (service member identity) are chosen for the three axes. The cube is built by selecting the attributes from the SERVICE_REPORT data mart shown in Table 6.3. The size and colour of the cube can be chosen to represent the number of the data records in the cube or the total values of a numeric attribute in those data records. The size of each individual cube in Figure 6.12 represents the number of records contained in it, whereas the colour of the cube indicates the total value of the attribute svc_repair_time for the records contained in the cube. The actual raw data for each cube can be viewed by right clicking the mouse.

For example, the cube selected in Figure 6.12 represents the one with the highest total service repair time (i.e. 19 days) for the machine model "MPA_G1_2547" by the service engineer "10404". The service time was 20 to 40 months before January 1997. As higher service repair time means longer machine down time, this may result in customer dissatisfaction. Since longer service time may be due to the inefficiency of the service engineer or the complexity of the machine model, the company should look into such problems in order to enhance the efficiency of service in the future.

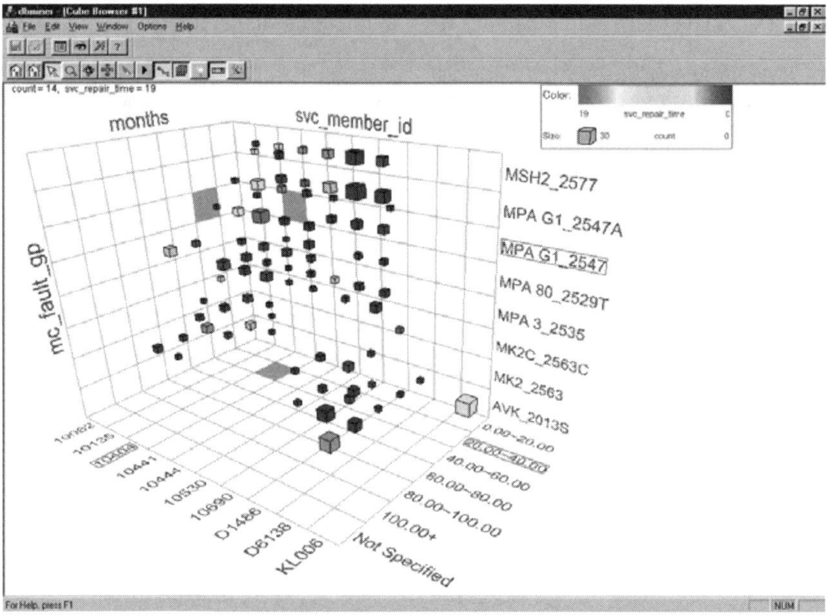

Figure 6.12 SVC_RPT_CUBE

After data cubes have been constructed, the next phase is to apply data mining functions to extract useful information. There are five major data mining functions: *Summarization*, *Association*, *Classification*, *Prediction* and *Clustering*.

6.4.7.1 Summarization

Data summarization characterizes a set of task-relevant data set by data generalization. Data summarization can be used with OLAP operations such as drilling, which allows users to examine data characteristics at multiple levels of abstraction. Data summarization can be implemented efficiently using OLAP operations on the data cubes. In addition, it enables dimension relevance analysis automatically in order to ascertain the selection of dimensions and the extent of the level of generalization. It summarizes data in the data cube at a desired level of generalization and presents the results using bar charts, pie charts, 3D bar charts and XY plots.

Figure 6.13 presents a summary of the machine models serviced by service engineers in the `SVC_RPT_CUBE` using a bar chart. This summarization is very useful for understanding the expertise of respective service engineers. This can be used to assign appropriate service engineers to service particular machine models. From the figure, it is evident that the machine model "AVK_2013S" has been serviced only by the service engineer "KL006". This probably means that "KL006" has good knowledge of this particular machine model. However, this also shows that the service engineer "KL006" has not worked on any other machine models.

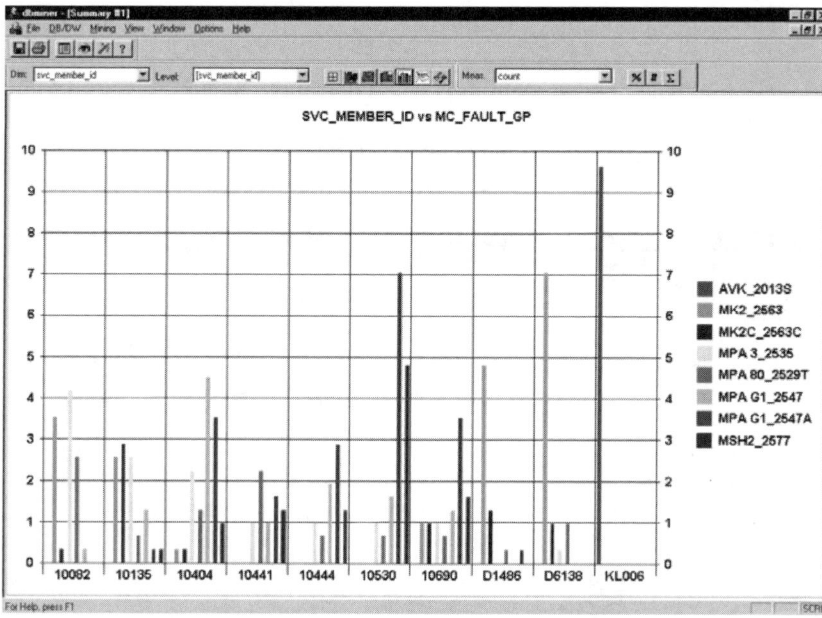

Figure 6.13 Summarization based on SVC_RPT_CUBE

6.4.7.2 Association

Application of Association generates rules [19] in the form $X \Rightarrow Y$ (*c*, *s*), which means "if a pattern *X* appears in a transaction, there is *c*% possibility (*confidence*) that the pattern *Y* holds in the same transaction and *s*% of total transactions contain (*support*) $X \cup Y$". The aim is to discover strong association rules (rules with high support and confidence) in large databases. Data cube-based association rule mining results in great flexibility and efficiency for three reasons. First, the data cube structure provides flexibility in grouping the data to a set of attributes. Second, the count and aggregate values in a data cube make the calculation of support and confidence in association rules easier, thereby enabling association testing and filtering. Third, drilling operations facilitate the mining of multi-level association rules.

Association rules are often viewed in textual format because this form of representation specifies the conditions and conclusions precisely, and it is possible to represent associations with more than two attributes. Figure 6.14 shows the association rules among the attributes of `SVC_RPT_CUBE` in textual format. The rules have a minimum support of 8% and a minimum confidence of 98%. High confidence rules represent distinctive patterns within a database. The association rules in Figure 6.14 provide much useful information such as:

- The machine models which have been repaired only by a particular service engineer.
- The service engineers who repair the machine faults efficiently (within a day).
- The machine models which have experienced multiple faults recently.
- The customers who have reported multiple fault problems recently.

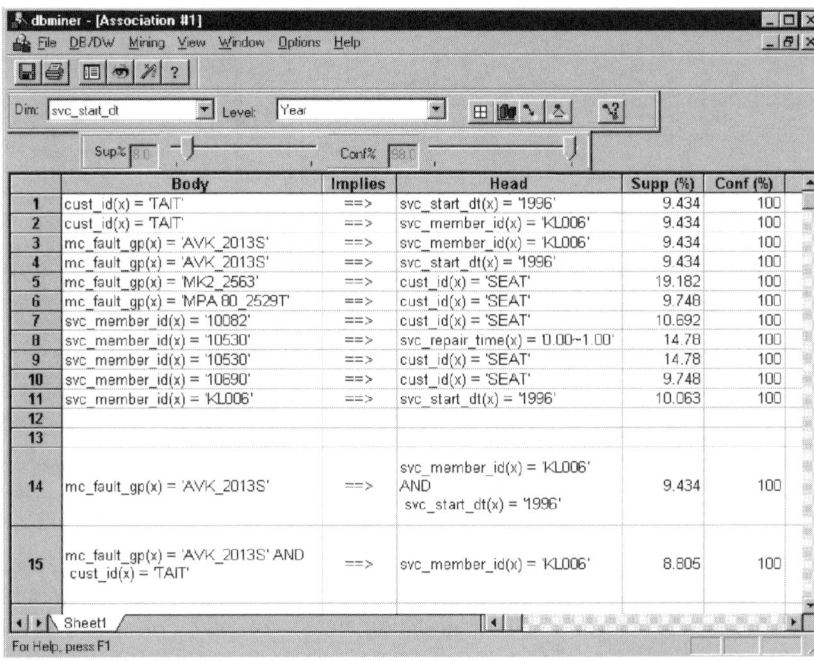

	Body	Implies	Head	Supp (%)	Conf (%)
1	cust_id(x) = 'TAIT'	==>	svc_start_dt(x) = '1996'	9.434	100
2	cust_id(x) = 'TAIT'	==>	svc_member_id(x) = 'KL006'	9.434	100
3	mc_fault_gp(x) = 'AVK_2013S'	==>	svc_member_id(x) = 'KL006'	9.434	100
4	mc_fault_gp(x) = 'AVK_2013S'	==>	svc_start_dt(x) = '1996'	9.434	100
5	mc_fault_gp(x) = 'MK2_2563'	==>	cust_id(x) = 'SEAT'	19.182	100
6	mc_fault_gp(x) = 'MPA 80_2529T'	==>	cust_id(x) = 'SEAT'	9.748	100
7	svc_member_id(x) = '10082'	==>	cust_id(x) = 'SEAT'	10.692	100
8	svc_member_id(x) = '10530'	==>	svc_repair_time(x) = '0.00~1.00'	14.78	100
9	svc_member_id(x) = '10530'	==>	cust_id(x) = 'SEAT'	14.78	100
10	svc_member_id(x) = '10690'	==>	cust_id(x) = 'SEAT'	9.748	100
11	svc_member_id(x) = 'KL006'	==>	svc_start_dt(x) = '1996'	10.063	100
12					
13					
14	mc_fault_gp(x) = 'AVK_2013S'	==>	svc_member_id(x) = 'KL006' AND svc_start_dt(x) = '1996'	9.434	100
15	mc_fault_gp(x) = 'AVK_2013S' AND cust_id(x) = 'TAIT'	==>	svc_member_id(x) = 'KL006'	8.805	100

Figure 6.14 Textual representation of Association rules for the SVC_RPT_CUBE

The first two association rules in Figure 6.14 state that customer "TAIT" had reported all the faults during the year "1996" and it was served by the service engineer "KL006" only. This observation can lead to the conclusion that service engineer "KL006" would be the most suitable person to be assigned to serve "TAIT" for future problems. The next two rules show that the machine model "AVK_2013S" is serviced by service engineer "KL006" only and the faults were reported in the year "1996". In addition, it can be seen that in rule 8, service engineer "10530" had resolved all the fault problems within a day.

6.4.7.3 Classification

Classification maps a data item into one of several predefined classes. It uses database records that reflect historical data to generate a model that analysts can use to classify new data, which is useful in determining target markets, for instance. Data mining goals 3 and 6 suggest that the company would like to know which machine models have experienced most faults recently and which customers have reported most of the faults.

6.4.7.4 Prediction

Prediction is similar to classification, except that it classifies attributes according to predefined future behaviour or estimated values. For example, suppose classification has indicated that, for model MPAG1_26547A, customer EPSB reported all the faults in 2000. Prediction might indicate that the same is likely to happen in 2003. Prediction capability is limited to the ranges of numerical data or categorical data that analysts can generalize to a small number of categories.

6.4.7.5 Clustering

Clustering partitions a data set into groups, or clusters, where all members of a particular cluster have some significant common basis. Unlike the classes in classification, clusters are based on the results of applying similarity metrics or probability density models. Clusters are displayed by showing the data cubes in 3D space. However, in the 3D display, only the clusters involving three attributes, at the most, can be shown.

Figure 6.15 shows the clusters formed with the combination of attributes `cust_id`, `mc_fault_gp` and `svc_member_id`. As shown, most of the data can be seen to be concentrated within a few clusters. For example, there are many instances which the service engineer "KL006" has served the customer "TAIT" for the machine model "AVK_2013S".

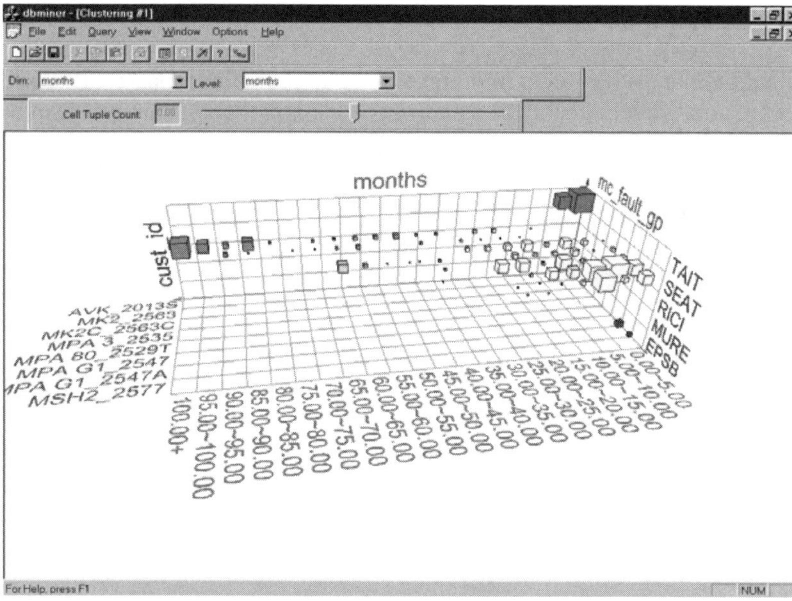

Figure 6.15 Clusters in the SVC_RPT_CUBE

6.4.8 Evaluate Mining Results

The final step is to gauge how well the data mining activities have helped us discover useful knowledge by comparing the results with our established goals:

- *Mining Goal 1* – OLAP analysis identifies the machine models with poor sales and higher frequency of faults. Summarization also identifies years when poor sales occurred.
- *Mining Goal 2* – Three-dimensional clusters have been formed using the clustering module based on machine models, their customers and the time of purchase. This provides valuable information for cross sales campaigns to target customers using other machine models.
- *Mining Goal 3* – OLAP and association can identify the machine models with a high frequency of faults. Classification and clustering also identify the customers who have reported the faults.
- *Mining Goal 4* – Service repair time is an important parameter in identifying efficient service engineers. Hence, association rules analysis is used to identify the service engineers who repair faults within a day. Then, the various attributes affecting the prediction of the service repair time are analysed. Since the type of machine models serviced can also affect the service repair time, prediction analysis is used to compare different service engineers who have repaired machines under the same conditions.

- *Mining Goal 5* – Summarization is first used to show the machine models repaired by the service engineers using bar charts. Apart from displaying general information on the expertise of service engineers, the bar chart also reveals certain interesting patterns. For example, service engineers who have repaired only a particular machine model, and the machine models that are repaired only by a particular service engineer are also identified. Moreover, association and prediction provide additional information on the time taken for different machines.
- *Mining Goal 6* – Customers who have reported many faults recently need better service support. Such customers can be identified using association rules analysis. Combined with mining goal 3, the company can provide better service to these customers based on their geographical locations and machine models purchased.

We have also sought feedback from the company's service engineers and administrators of the customer service department. In general, customer service personnel saw the seven-step process as a way to gain in-depth understanding of the knowledge hidden in stored data. They found our process useful in providing better customer service that would help them remain competitive. Some felt that data mining represented a way to improve the performance of manufacturing processes as well.

6.5 WEB-BASED INTELLIGENT SURVEILLANCE SYSTEM

Security is a mounting concern, both for corporations as well as individuals due to the increasing threats from terrorists and other criminals. A Web-based intelligent security surveillance system is desirable because it is cost-effective and convenient (authorized users simply need to access the system via a Web browser), and can relieve the burden on users as they no longer need to keep a constant watch on the monitored sites. With recent technological advances, such a system can now be realized. This section describes the major technical challenges – and solutions – that have been involved in the development of a Web-based intelligent security surveillance system.

6.5.1 Design Objectives and Related Systems

A Web-based intelligent security surveillance system enables authorized users to remotely monitor multiple sites from afar, via the Internet. To be "intelligent", the system must analyse monitored scenes in real-time and alert the human operator of impending trouble so that actions could be taken. This can alleviate the burden on security personnel. The design objectives may be summarized as follows:

- Cost effective – The system should incur minimal startup and maintenance costs.

- User-friendly and convenient – The system should provide user-friendly features such as VCR-like operations for video playout and recording. The system should also be accessible by authorized personnel anytime, anywhere, via the WWW.
- "Intelligent" – Real-time interpretation of human activities in the monitored scenes.
- Scalable – Deployment should be easy regardless of the size of the system measured in terms of the number of users, the number of monitored sites, and so on.
- Good video playout quality – If suspicious activity is detected in a monitored scene, a signal is generated to alert the user, who might like to look at the video sequence before taking action.
- Secure – System security is of utmost importance.
- Modular components – Such flexibility would facilitate future upgrades when new technologies become available.
- The system should be better than existing (commercial) solutions.

Current systems that are most closely related to our work are mainly used for intrusion detection and access control. Two of the most representative systems are Video Motion Detection (VMD) [20] and Intelligent Scene Monitoring (ISM) systems [21]. VMD systems respond to temporal and/or spatial variations caused by movements between video frames. Video frames are first sampled at some intervals and stored as reference. When the same area is scanned, new video frames are compared with the reference frames. An alert is triggered if there are significant differences between them. The monitored scene is usually divided into a number of sections, each of which is monitored separately. Detection is based on monitoring all sections or a combination of the selected sections. Some sections are ignored where motion detection is not required or where false alarms can result due to natural phenomena such as cloud movements. The target size (e.g. to track human-size objects), target speed, and sensitivity (to minimize false alarms) are adjustable.

ISM systems attempt to interpret the scene rather than just observing passively like VMD systems. An alarm is raised only when a specific user-defined event (or sequence of events) occurs in the scene. Two processes are involved: detection (detect and segment moving objects or regions) and recognition (extract motion events of interest). While VMD systems have been successfully applied to such passive security operations as personnel access control, ISM systems further reduce the likelihood of false alarms due to rapid environmental changes by event analysis.

However, human body motion analysis [22] is required for the detection of a range of criminal activities. It is used to distinguish between "normal" motions and "criminal activities". The objective is to recognize human motion types such as walking, running and wrestling. Various approaches have been proposed for different phases of analysis: segmentation of a moving human body from the background and extraction of low-level features [23], body part location which includes moving parts tracking and labeling of the extracted regions [22] and motion pattern analysis [24].

6.5.2 System Overview and Major Components

As shown in Figure 6.16, the Web-based intelligent security surveillance system has four main components.

- *Monitor Node* – It performs video capture, digitization and analysis.
- *Monitoring Server* – It manages a number of Monitor Nodes within a geographic cluster and performs video recording/retrieval.
- *Exchange Server* – It connects between the Monitoring Servers and Clients.
- *Monitoring Client* – This is the user interface, which is invoked by launching a Web browser.

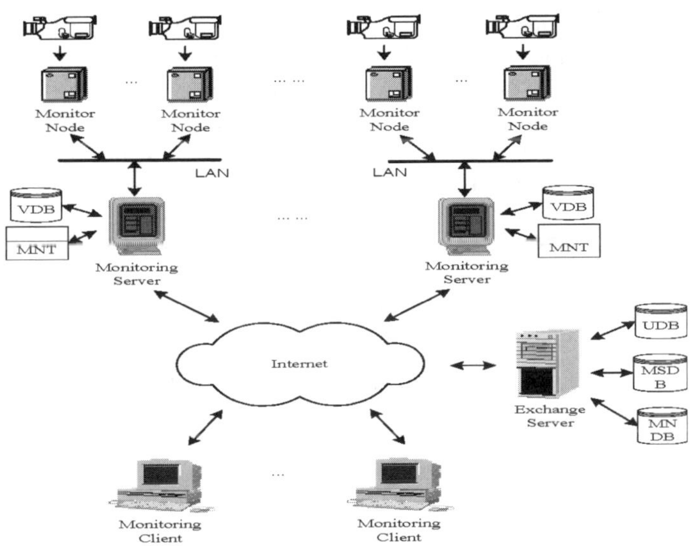

Figure 6.16 An overview of the system

6.5.2.1 Monitor Node

As shown in Figure 6.16, there are multiple Monitor Nodes typically installed in different locations. Each Monitor Node captures video in the monitored sites continuously and performs scenario analysis for suspicious events. If a suspicious event is detected, the Monitor Node generates an alert signal. Monitor Nodes that are located in close proximity with each other form a geographic cluster. The cluster is connected to a Monitoring Server that manages all Monitor Nodes in that cluster through a dedicated connection (LAN). A Monitor Node consists of four components: *Video Camera*, *Scenario Analyser, Request Parser* and *Video Manager*.

The video camera, which can be monochrome or colour, is used to capture and digitize the monitored scene. The Scenario Analyser uses an algorithm to detect suspicious activities in the captured video sequence.

This algorithm will be described in Section 6.5.4.5. The request parser handles requests such as asking for continuous monitoring from the Monitoring Server. It provides an interface for other components (especially the Monitoring Server) to communicate with the Monitor Node using a simple protocol.

It also performs authentication with the requesting system before proceeding to interpret a request. The request is interpreted and passed to the Video Manager, which responds to the request.

The video manager is a Monitor Node's main component. As shown in Figure 6.17, it performs request handling, video compression and buffering, alert generation (if suspicious activity is detected), and status refreshing (for example, when status changes from "fighting" to "normal"). This event-triggered status refreshing mechanism means that the Monitoring Server that handles this Monitor Node does not need to poll for latest status periodically.

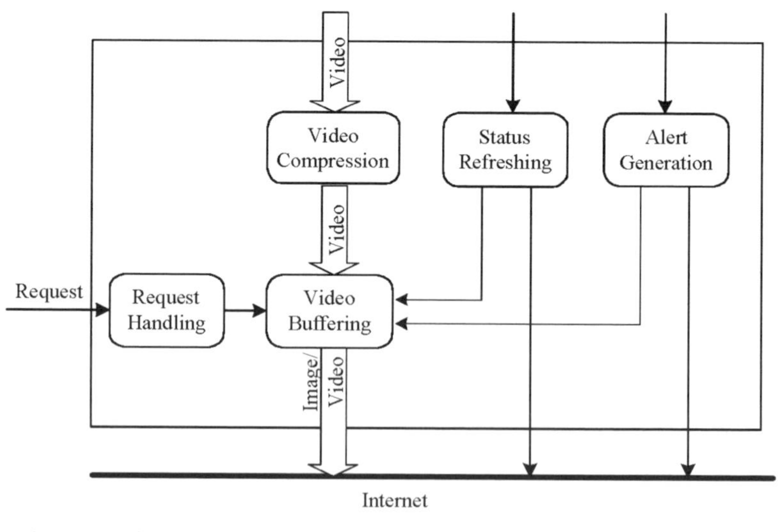

Figure 6.17 Monitor node's video manager

6.5.2.2 Monitoring Server

The Monitoring Server acts as a gateway between Monitoring Client(s) and Monitor Node(s). Each Monitoring Server handles a group of Monitor Nodes in the same neighbourhood. It maintains a Monitor Node Table (MNT) with latest information such as change of status from the Monitor Node(s). It is also responsible for the storage and retrieval of video sequence/images of the monitored sites. In addition, it relays any alert signals from the Monitor Nodes to all connected clients.

Figure 6.18 shows the schematic of a monitoring server. The *Request Parser* handles requests from the Monitoring Clients, and the *Request Generator* formulates the request to be sent to the appropriate Monitor Node. *Negotiation* handles the communications between the Monitoring Client(s) and Server. The *Alert Watchdog* maintains a dedicated TCP connection for incoming alerts from the monitor nodes. It updates the MNT and relays the alert to the connected client(s). Again, video buffering is needed because the video frames from the Monitor Nodes may arrive out of order. In addition, a video database is used to store video footage for subsequent retrieval.

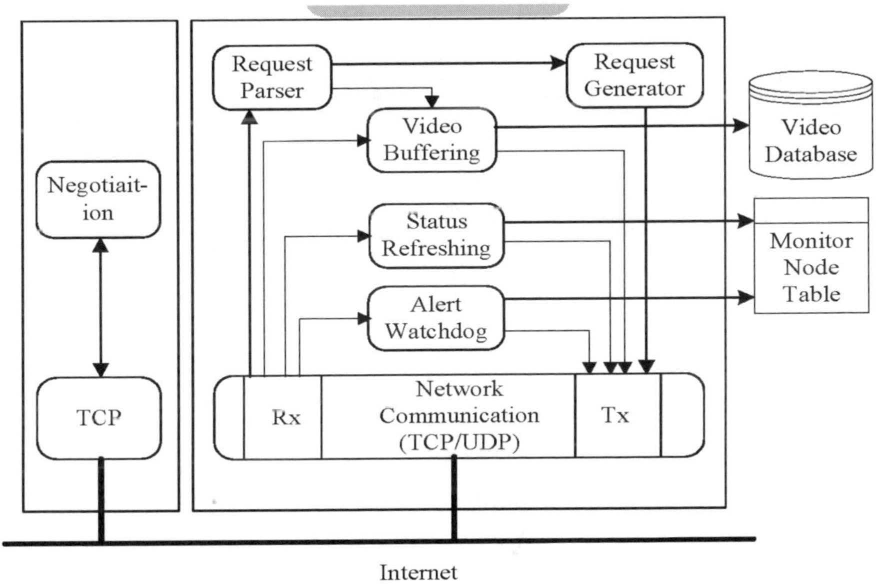

Figure 6.18 Monitoring server

6.5.2.3 Exchange Server

As shown in Figure 6.18, multiple Monitoring Servers are allowed to promote system scalability. However, it would be inconvenient for users to remember IP addresses of all Monitoring Servers in the system. The use of an Exchange Server simplifies the management of these Monitoring Servers. It maintains a Monitoring Server Database (MSDB) and a Monitor Node Database (MNDB) to keep track of all Monitoring Servers and Monitor Nodes in the system. An extra advantage is that addition and removal of any Monitoring Server from the network will not directly affect the clients.

Figure 6.19 highlights the role played by the Exchange Server. First, the user connects to a Web-based Monitoring Client and requests for the list of all Monitoring Servers that the user is authorized to access. Next, the Exchange Server sends the list to the user in response to the request. Finally, communication is established directly between the user (via the Monitoring Client) and the selected Monitoring Server.

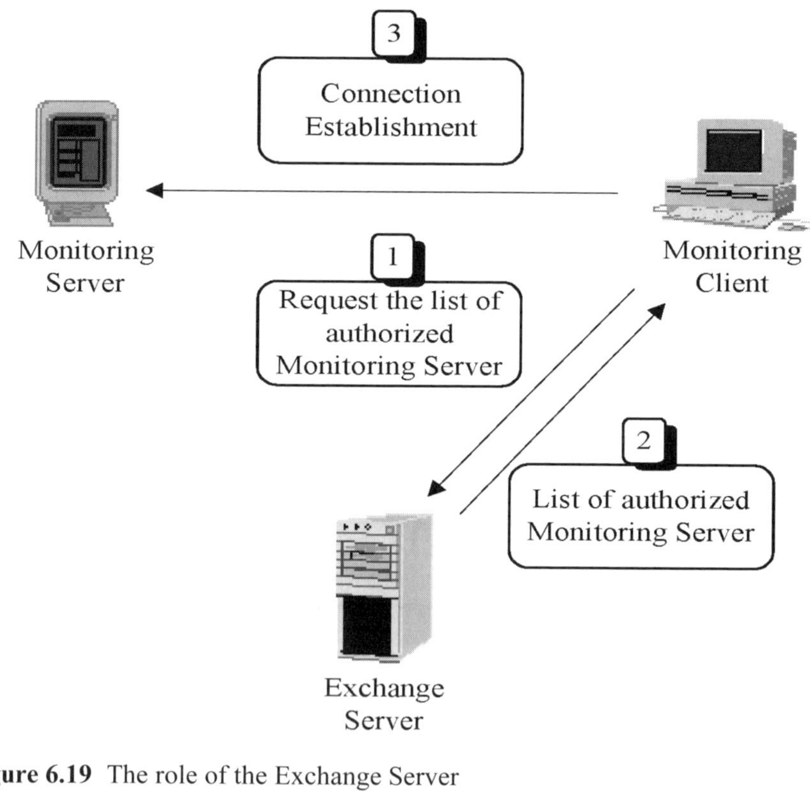

Figure 6.19 The role of the Exchange Server

6.5.2.4 Monitoring Client

The Monitoring Client is shown in Figure 6.20. Invoked by launching a Web browser, it acts as an interface between the user and the system. It connects to the Exchange Server to request a list of all Monitoring Servers that the current user is authorized to access. Upon successful connection to the selected Monitoring Server, information about all Monitor Nodes handled by the selected Monitoring Server is retrieved and forwarded to the client. The Monitoring Client also provides an interface for system administration.

Figure 6.20 The Web-based Monitoring Client window

6.5.3 Monitoring Process

The monitoring process is illustrated in Figure 6.21. As shown in the figure, there are four main phases:

- *Software Activation* – The user invokes a Monitoring Client by launching a Web browser and logging onto the system. The system is currently password protected so that only authorized users can access it. The username/password is encrypted and sent to the Exchange Server for verification.

- *Monitoring Server Resolution* – The Exchange Server performs user verification. Upon receiving a request from the user via the Client, the Exchange Server responds by sending a list of all Monitoring Servers that the current user is authorized to access. Once the user has selected the preferred Monitoring Server, information about all the Monitoring Server's Monitor Nodes is sent to the Client.
- *Online Monitoring* – Three events can take place: status refreshing (whenever there is a change in status of any of the active Monitor Nodes), alert signals (generated whenever suspicious activity is detected) and continuous monitoring (if the user chooses to keep constant watch)
- *Connection Termination* – This is activated when the user logs off the system.

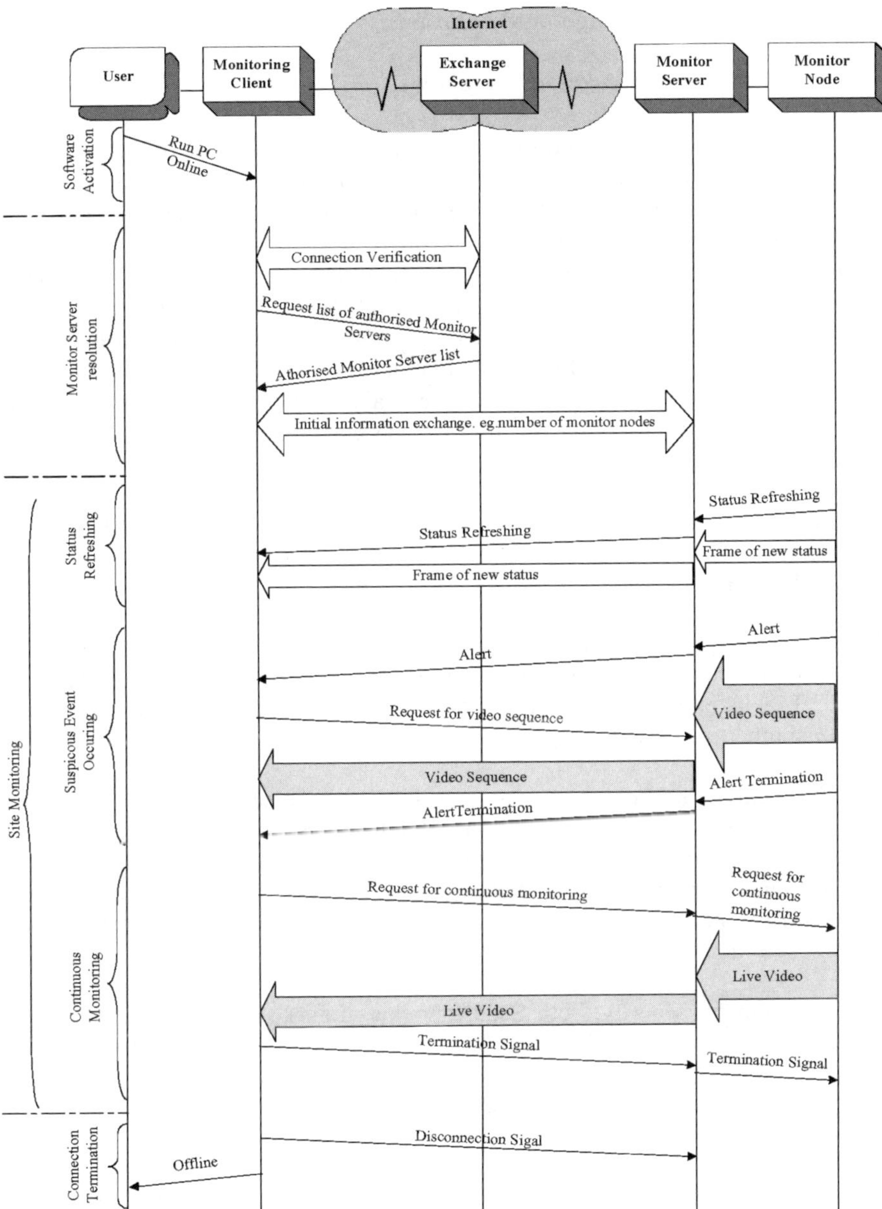

Figure 6.21 The monitoring process

6.5.4 Technical Challenges and Solutions

The major technical challenges are as follows:

- *Security* – This is of utmost importance due to the nature of this system.
- *Compression standards* – The video data must be suitably compressed before transmission to minimize the bandwidth requirements. Different compression standards are investigated.
- *Internet Communications Protocols* – Different protocols (e.g. TCP, UDP, etc.) are used to support different types of data transmitted.
- *Quality of Service (QoS) control mechanisms* – It is important to provide an acceptable QoS of the video playout.
- *Video sequence analysis* – It is necessary to interpret the scene being captured on video so that an alert can be generated if suspicious activity is detected.

These challenges – and solutions – are discussed in the following subsections.

6.5.4.1 Security

There are two aspects of security of concern: *access security* and *transfer security*. The former refers to the process of authenticating users trying to gain access to the system. The latter refers to the protection of data that are transferred from one point to another over the network.

To deal with access security, our initial implementation uses encrypted username/password that is checked by the Exchange Server. Currently, we are developing a biometric system that uses multi-view facial analysis for authentication [25]. Although, we have found the encrypted username/password approach generally adequate, the biometric system can be employed for access control if extra security is needed.

For transfer security, both Secure Sockets Layer (SSL) and Digital Signatures are not suitable for video data. SSL is mainly for TCP, which is not suitable for transfer of video data. Using Digital Signatures would mean inserting these special bit sequences into the video bit stream, which would incur a great deal of additional processing effort.

We have, therefore, adopted a hybrid encryption scheme as follows. Although public-key encryption is slow compared to private-key encryption, it avoids the key distribution problem associated with the latter. A hybrid system that combines the use of both methods offers good compromise in terms of security and speed of operation. In this scenario, public-key encryption is first used to pass a secure one-time-use private key to both communicating parties. They may then exchange data using private-key encryption.

6.5.4.2 Compression Standards

Compression is necessary for video data due to the very large bandwidth requirements. Motion JPEG is used in the original design due to better data protection – it only uses intra-mode compression. This effectively amounts to switching off inter-mode coding (as in MPEG) to minimize the effects of error propagation. Further work focuses on MPEG4 for object-based interactivity.

6.5.4.3 Internet Communications Protocols

HyperText transfer Protocol (HTTP) over Transmission Control Protocol (TCP) is only used to transfer control information about the real-time data between client and server. Information transmitted in this way may include video data encoding type (MPEG video or Motion JPEG video, etc.), or the Monitoring Server's address for retrieving the required video data. Real-time Streaming Protocol (RTSP) is used for the exchange of control information required to start and stop data transmission. It supports VCR-like features: start, stop, fast-forward and pause. Finally, Real-time Transport Protocol (RTP) over User Datagram Protocol (UDP) is used for the actual video transmission.

6.5.4.4 QoS Control for Video Transmission

We have employed a variation of the IQCM scheme to ensure that an acceptable playout quality is delivered to the user. Compared to other video applications such as video on demand (VOD) for movies, the frame rate required for security surveillance can be much lower. We have successfully tested our system at frame rates of between 1 and 3 frames per second (fps). The system has been found to be effective at such low frame rates. Figure 6.22 summarizes the schematic of the QoS control mechanism that we have used.

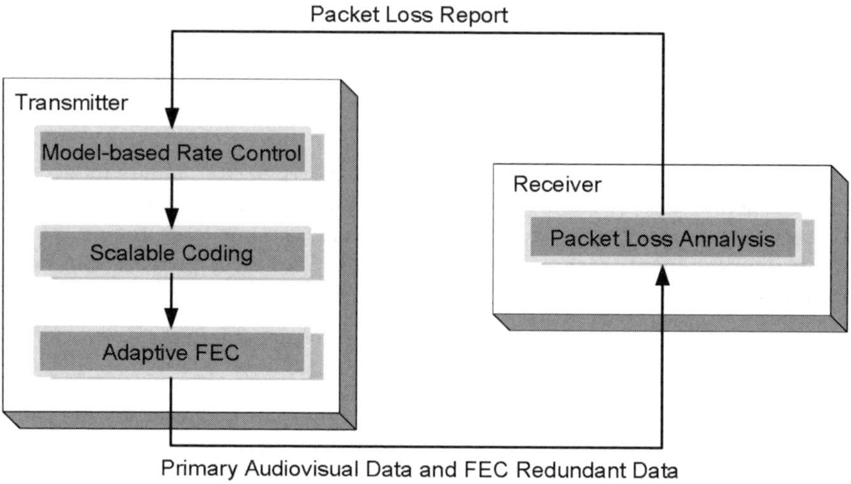

Figure 6.22 QoS control mechanism for the Web-based security surveillance system

As shown in Figure 6.22, the QoS control mechanism used is both integrated and adaptive. The main features of this mechanism are:

- Rate-based Rate Control performed by the sender, network condition modelling by each Receiver
- Integrated Rate-Adaptive Coding and Rate Shaping based on multiple (scalable) bit streams
- Adaptive Forward Error Correction (FEC) that varies number of redundant packets sent with the prime data

6.5.4.5 Video Sequence Analysis

It would be good to make machines analyse and interpret video scenes. Despite years of development, however, no generic, fully automated techniques have emerged to tackle the problem. Currently, most of the best solutions are semi-automated, which require some user input. There are three main phases involved in video sequence analysis [26]:

- *Video Object Segmentation* – Semantically meaningful objects (regions of interest) are extracted from the video sequence.
- *Feature Extraction* – Extract information about the segmented objects.
- *Scenario Analysis* – Events are analysed and classified into one of the predefined possibilities.

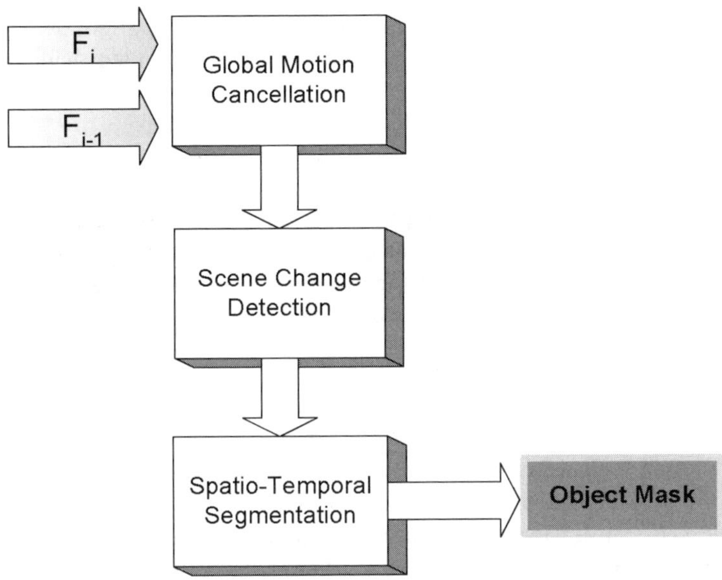

Figure 6.23 Video object segmentation

Figure 6.23 summarizes the steps involved in video object segmentation. Two consecutive F_i and F_{i+1} are taken as inputs. First, Global Motion Cancellation is used to cancel the effects of camera movement in relation to the scene being captured. In situations where the camera is fixed relative to the scene being captured (e.g. fixed security camera installed in confined spaces such as elevators/lifts, banks, etc.), Global Motion Cancellation is not required and this step can be omitted (switched off).

Scene Change Detection is necessary because temporal segmentation can become unreliable if a scene change has occurred between the two consecutive frames under consideration. Scene change analysis can be performed by histogram analysis: if the histograms of frames F_i and F_{i+1} are significantly different, then a scene change has occurred, and temporal segmentation should be switched off momentarily. Scene Change Detection can also be accomplished more simply by comparing the mean absolute difference between frames F_i and F_{i+1}: if the difference is greater than a preset threshold, then scene change has occurred.

For segmentation of semantically meaningful video objects, both spatial and temporal segmentation techniques are needed because none of them is sufficient on its own. Spatial segmentation is applied to individual frames to divide the entire 2D frame into homogeneous regions in terms of low-level properties such as intensities. This is accomplished as follows. First, unnecessary details such as texture information are removed using morphological filters. In particular, the morphological opening operator $\gamma(F_i)$ is used to remove unwanted highlight details logical opening operator $\gamma(F_i)$ is used to remove unwanted highlight details and the morphological closing operator $\varphi(F_i)$, is used to remove unwanted shadow details.

$$\gamma\,(F_i\,(x, y)) = \delta\,(\,\varepsilon\,(F_i\,(x, y))\,) \tag{6.1}$$

$$\varphi\,(F_i\,(x, y)) = \varepsilon\,(\delta\,(F_i\,(x, y))\,) \tag{6.2}$$

The $\gamma(F_i)$ operation is essentially a morphological erosion $\varepsilon(F_i)$ followed by a morphological dilation $\delta(F_i)$, whereas $\varphi(F_i)$ is the reverse operation. In all cases, a flat structuring element is used for the morphological operations. $\delta(F_i)$ and $\varepsilon(F_i)$ are in turn defined as follows for all (p, q) in the region of interest

$$\varepsilon\,(F_i\,(x, y)) = min\,(F_i\,(x + p, y + q)) \tag{6.3}$$

$$\delta\,(F_i\,(x, y)) = max\,(F_i\,(x - p, y - q)) \tag{6.4}$$

A morphological gradient operator is then used to extract region boundary information. The morphological gradient is obtained simply by subtracting $\varepsilon(F_i\,(x,y))$ from $\delta(F_i\,(x,y))$. Morphological gradients obtained using symmetrical structuring elements are preferable to derivative gradients as they depend less on edge directionality than the latter. Next, the local gradient minima in each F_i are used as seeds for region growing. Region growing is generally performed by mathematical morphology. In this research, the watershed algorithm [27] is adopted for region growing. However, region growing methods tend to produce over-segmented frames due to gradient noise. The result is that, often, there are isolated regions that should be connected to form a semantically meaningful object. Thus, it is necessary to apply region merging to the over-segmented frame. This is performed by a combination of similarity measure relaxation and morphological erosion.

The last step in the spatial approach is to remove noise within a frame. Small regions are treated as noise and removed. The small region filtering process comprises a region labelling process and a small region removal process. In this research, any region with an area smaller than 30 pixels is regarded as noise and is therefore discarded.

From a temporal perspective, motion is the most important cue for segmentation. Since moving objects in video sequences may be observed as changes in intensity, it is possible to identify moving objects by considering changes in intensities between consecutive frames. Often, there are intensity differences between the supposedly stationary backgrounds captured in two consecutive frames due to noise. The noise can be filtered out as it can be statistically modelled as a normal distribution [28]. Temporal segmentation of video objects begins with the isolation of the background.

For most commercial or industrial applications, the monitored sites are typically confined or semi-confined spaces, such as a high-security factory. This allows us to use a simple subtraction technique for temporal segmentation. In particular, we apply two pixel-wise intensity differences to separate the moving objects based on inter-frame motion and motion against the stationary background. These differences are defined by the following, for all x, y, i

$$Diff_{\text{inter-frames}}(F_i) = F_i(x, y) - F_{i-1}(x, y) \qquad (6.5)$$

$$Diff_{\text{background}}(F_i) = F_i(x, y) - F_0(x, y) \qquad (6.6)$$

where $F_0(x,y)$ pre-captured frame of background

After subtraction, a thresholding operation is needed to completely separate the target regions of interest from the background. The histogram of the difference pictures is analysed to provide an appropriate intensity threshold. Due to the effects of noise, however, the subtraction does not always give clear-cut solutions. Thus, pixels with intensity differences less than a predefined threshold are treated as background and are discarded.

The purpose of Feature Extraction is to obtain as much information as possible about the various objects of interest. For example, it is sometimes necessary to identify unusual stationary objects, such as graffiti or bloodstains. Most commonly, however, obtaining information from moving objects is necessary for the detection of suspicious activities. First, it is useful to estimate the heights of (human) objects for example, children or adults, standing or lying on floor, and so on. Next, these objects of interest should be tracked across the video frames and with the speed of the objects' movement, estimated to determine likely activities.

Scenario representation [29] is used to represent a human activity defined by a sequence of events. An event can be considered as a snapshot of the activity at any instant moment. It is characterized by enabling features that describe the event. Enabling features for an event are required to portray the spatial changes in motion configuration of the event. The temporal conditions may include the occurrence order of events, starting point, duration, tempo and intervals between consecutive events. The occurrence order of events specifies the chronological relationships between events of an activity. Starting point records the start time of occurrence. Duration of an event keeps a record of the minimal and maximal time span. Tempo and intervals then describe the occurring frequency of the events and ranges of time between consecutive events.

We employ a bi-directional search technique for scenario analysis as shown in Figure 6.24. It is used because it effectively restricts the search space, which in turn minimizes processing time. The algorithm takes background context information, such as room and lighting information, and status information, such as windows/doors closed/open, as inputs. Such input information may be derived from background reference frames and/or other sensors (e.g. a simple binary switch for door open/closed status). Relevant information from the feature extraction stage is used to trigger a new search to determine the latest scenario.

From the Feature Extraction stage, a bottom-up feature collection and analysis process is followed by a top-down verification process. Using user-defined and context-dependent threshold values, if the collected feature information agrees with a scenario, then a conclusion is drawn. It is important to note that it is not always necessary to receive input from all sources in order to conclude that a certain scenario is detected. In further research, we are developing a more sophisticated video sequence categorization engine that uses more features, as well as the learning capabilities of neural networks, for analysis.

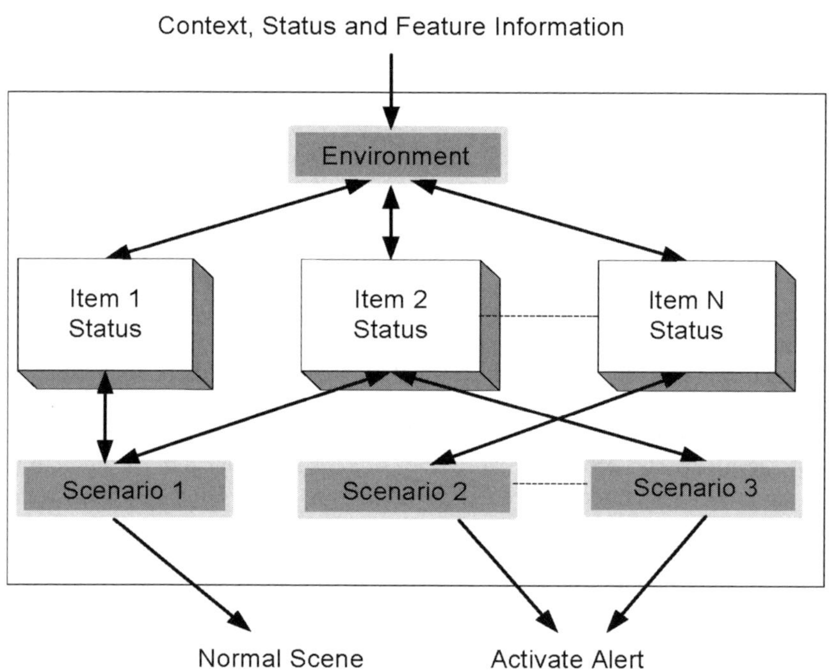

Figure 6.24 Bi-directional search for scenario analysis

We have implemented an experimental system to test the scenario analyser connected to security cameras already installed in elevators/lifts. As an example, Figure 6.25 illustrates three frames selected from a video sequence used in the test in which fighting occurs between two occupants. Each frame is indexed by (video sequence number: frame number). The frame rate was 2 fps. Suspicious activity due to excessive movement was successfully detected at frame 3:16. The elevator door was closed throughout the period when excessive movement was detected.

Based on head top points analysis, the system further detected that one of the occupants had fallen. An alert signal was activated in this sequence.

(3:12) (3:16) (3:22)

Figure 6.25 Suspicious activity was detected in test sequence 3

References, Links and Bibliography

[1] C-H Wu, J.D. Irwin and F.F. Dai, "Enabling multimedia applications for factory automation", *IEEE Trans. Industrial Elect.*, Vol. 48, No. 5, pp. 913–919, Oct 2001

[2] The OBI Consortium, Open Buying on the Internet http://www.openbuy.org/obi/library/white-paper.html

[3] M. Mori, H. Tsuru, R. Itsuki, H. Kitajima and H. Yajima, "Proposal of application architecture in electronic commerce service between companies", *Advance Issues of E-Commerce and Web-Based Information Systems, WECWIS*, 46–49, 1999

[4] http://www.cisco.com

[5] http://www.dell.com

[6] http://www.gte.com

[7] http://www.tradeout.com

[8] http://www.chemdex.com/index.html

[9] http://www.esteel.com

[10] Z. Tian, L.Y. Liu, J. Li, J.Y. Chung and V. Guttemukkala, "Business-to-business e-commerce with open buying on the internet", *IEEE Advance Issues of E-Commerce and Web-Based Information Systems, WECWIS*, 56–62, 1999

[11] P. Dasgupta, N. Narasimhan, L.E. Moser. and P.M. Melliar-Smith, "MAgNET: Mobile agents for the networked electronic trading", *IEEE Trans. Knowledge and Data Engineering*, Vol. 11, No. 4, pp. 509-525, 1999

[12] S. Foo, S.C. Hui and S.W.Yip, "Enhancing the quality of low bit-rate real-time Internet communication services", *Internet Research: Electronic Networking Applications and Policy*, Vol. 9, No. 3, 212–224, 1999

[13] T. Kohonen, *Self-Organizing Maps*, Second Extended Edition, Springer Series in Information Sciences, Vol. 30, Springer, Berlin, Heidelberg, New York, 1997

[14] http://www.jsc.nasa.gov/~clips/CLIPS.html

[15] D.W.R. Patterson and J.G. Hughes, "Case-based reasoning for fault diagnosis", *The New Review of Applied Expert Systems*, Vol. 3, pp. 15–26, 1997

[16] A.C.M. Fong and S.C. Hui, "An intelligent online machine fault diagnosis system", *Computing & Control Engineering Journal*, Vol. 12, No. 5, pp. 217–223, 2001
[17] C. Fellbaum (ed.). *Wordnet: An Electronic Lexical Database*, The MIT Press, 1998
[18] A.C.M. Fong, S.C. Hui and G. Jha, "Data mining for decision support", IEEE *IT Professional*, Vol. 4, No. 2, pp. 9–17, 2002
[19] J. Han and Y. Fu. "Discovery of multi-level association rules from large databases". *Proc. Int. Conf. Very Large Data Bases (VLDB'95)*, pp. 420–431, Zurich, Switzerland, Sept. 1995
[20] R.M. Rodger, I.J. Grist and G.A.O. Peskett, "Video Motion Detection Systems: A Review for the Nineties", *International Carnahan Conference on Security Technology*, pp. 92–97, 1994
[21] R.F. Mackintosh, "Sentinel: A Technology for Practical Automatic Monitoring of Busy and Complex Scenes", *International Carnahan Conference on Security Technology: Crime Countermeasures*, pp. 190–193, 1992
[22] M.K. Leung and Y.H. Yang, "First Sight: A Human Body Outline Labeling System", *IEEE Trans. Pattern Analysis and Machine Intelligence*, Vol. 17, No. 4, pp. 359–377, 1995
[23] S.N. Jayaramamuthy and R. Jain, "An approach to the segmentation of textured dynamic scenes", *Computer Vision, Graphics and Image Processing*, Vol. 21, pp. 239–361, 1983
[24] G. Ferrigno, N.A. Borghese and A. Pedotti, "Pattern recognition in 3D automatic human motion analysis", *ISPRS Journal of Photogrammetry and Resensing*, Vol. 45, pp. 227–24, 1996
[25] Y Gao, S.C. Hui and A.C.M. Fong, "A multi-view facial analysis technique for identity authentication", *IEEE Pervasive Computing*, Vol. 2, No. 1, pp. 38–45, 2003
[26] A.C.M. Fong, S.C. Hui and M.K.H. Leung, "Content-based video sequence interpretation", *IEEE Trans. Consumer Electronics*, Vol. 47, No. 4, pp. 873–879, 2001
[27] L. Najman and M. Schmitt, "Geodesic saliency of watershed contours and hierarchical segmentation", *IEEE Trans. Pattern Analysis Machine Intell.*, Vol. 18, No. 12, pp. 1163–1210, 1994
[28] Y.Z. Hsu, H.H. Nagel and G. Rekers, "New likelihood test methods for change detection in image sequences", *Comp Vision, Graphics, Image Processing*, Vol. 26, pp. 73–106, 1984
[29] E. Rich, "Artificial Intelligence", McGraw-Hill, pp. 201–242, 1983

CHAPTER 7

IMPLEMENTING AND DELIVERING INTERNET AND MULTIMEDIA PROJECTS

7.1 INTRODUCTION

The World Wide Web provides an interesting medium for presenting individuals with a personal homepage, selling business products via an online shop, and gathering and distributing information to/from vast audiences at the click of the mouse. A company can make a large amount of information readily available to potential buyers at a low cost on the web. In fact, commercial web sites and Internet access accounts are growing at an unbelievable rate and becoming the fastest growing segment of the industry.

Web developers are facing growing technical options, from programming languages to page designs to the latest and flashiest technologies. Due to fierce competition, it remains challenging to produce competitive business applications on the Internet, particularly with the increasing amount of multimedia applications dominating the market. Developing Web content entails shaping and negotiating, and making many choices involving technical, aesthetic and usability concerns. Web developers today need a more process-oriented approach in order to articulate the information content they wish to convey through the Internet.

Fortunately, Web developers can draw on many existing concepts from software engineering practices. This chapter introduces applicable software engineering methodologies for developing Internet and multimedia business applications. We initially talk about the methodologies of developing Web-oriented applications. Then we present the common Web development processes we use to produce Internet-based products, from planning, analysis, design and implementation. Finally, we discuss the important issue of measuring and managing the quality of the Web-based business applications. In particular, we consider:

- Process modelling and lifecycle.
- Project planning and management.

- Design, implementation and testing.
- Measurements.

7.2 PROCESS MODELLING AND LIFECYCLE

The development of software systems has become more and more complex. A common software development process includes many milestones, deliverables and checkpoints. In modern software engineering, a software life cycle model is commonly used to depict the significant phases or activities of a software project from conception until the product is retired. A typical life cycle model includes the following phases of a software project: requirements analysis, design, coding and unit testing, system integration, system testing, operations and maintenance. There are several common life cycle models that are suitable for Internet-based software development: waterfall model, spiral model, prototyping model and incremental and iterative development model. Since the life cycle steps are described in very general terms, the models are adaptable and their implementation details will vary among different organizations. Organizations may mix and match different life cycle models to develop a model tailored to their products and needs.

7.2.1 Waterfall Model

The history of Software development life cycle models can be dated back towards the 1970s, when Royce first publicly documented the "waterfall model" [1]. As shown in Figure 7.1, the waterfall model depicts the development stages as cascading from one to another, which means one development stage should be completed before the next begin. Today, the Waterfall model is the least flexible and most obsolete of the life cycle models. It is most suitable to projects that have low risk in the areas of user interface and performance requirements, but high risk in budget and schedule predictability.

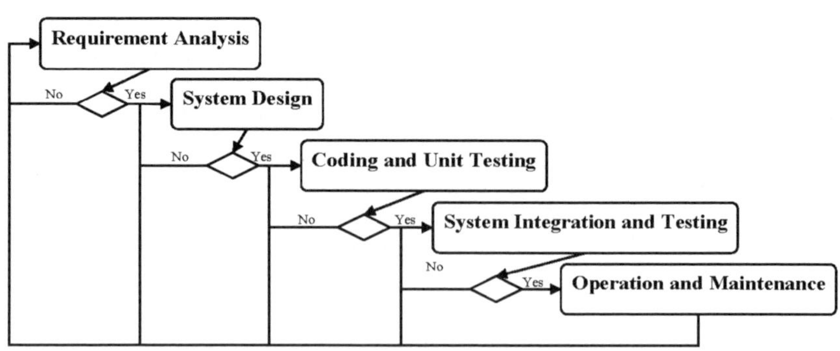

Figure 7.1 Waterfall model

7.2.2 Spiral Model

Addressing the ever-increasing risks in software project management, Boehm in 1985 suggested that the development lifecycle could be viewed as "spiral model" [2]. As shown in Figure 7.2, the spiral model begins a development cycle with a planning phase to define resources, timelines and other project information. The model then inserts a risk analysis step to evaluate both technical and management risks and even producing a prototype to make a "go" or "stop" decision before further engineering and construction effort is put in. The engineering and construction phase builds and tests one or more representations of the application. The results are passed on to the customers for evaluation. The spiral incorporates prototyping as a risk reduction strategy. It is the most generic of the models. Most life cycle models can be derived as special cases of the spiral model.

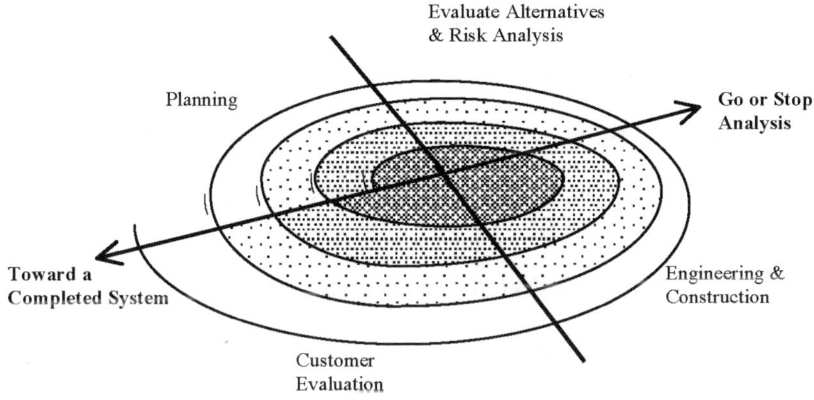

Figure 7.2 Spiral model

7.2.3 Prototyping Model

The prototype model is illustrated in Figure 7.3. It begins with requirements gathering. The project team and the customer meet together to define the overall objectives of the prototype. Then the development team design and build a prototype quickly.

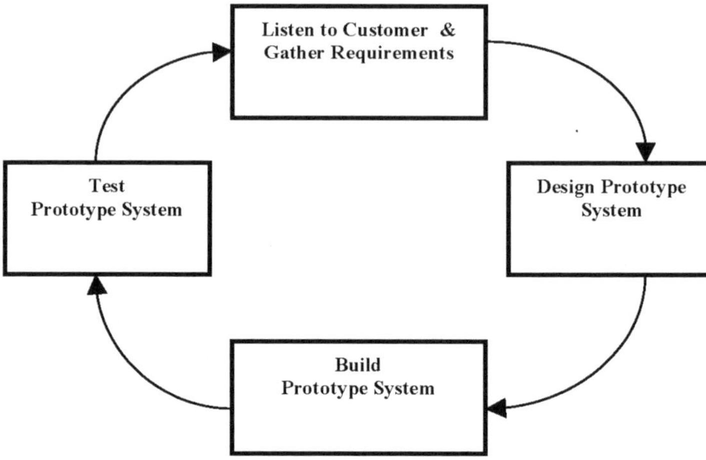

Figure. 7.3 Prototyping model

The prototype is then evaluated by the customer to get feedback and to refine the requirements. Iteration occurs as the prototype is evolved to satisfy the needs of the customer. The prototype model is best used in projects that have high risk in user interface design.

7.2.4 Incremental and Iterative Development

The incremental model delivers software in small but usable pieces, called “increments”. In general, each increment builds on those that have already been delivered [3]. Each increment constructs several partial deliverables, each having incrementally more functionality than the previous increment. Unlike the iterative model, the incremental model focuses on the delivery of an operational product with each increment. In contrast, the iterative model builds a full system from the start but only with partial functionality. In each iterative cycle, more functions are added until a complete system is built. This is illustrated in Figure 7.4.

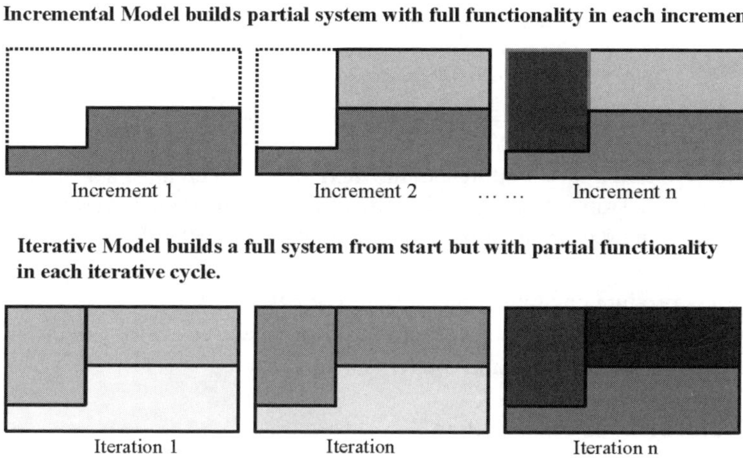

Figure 7.4 Incremental and iterative model

7.3 PROJECT PLANNING AND MANAGEMENT

With the emergence of Multimedia technology in the early 1990s, numerous interactive multimedia systems have been developed in areas such as news delivery and entertainment, training and education, advertising and broadcasting as well as online trading. A multimedia system designer is no longer a "programmer", but a software engineer who uses software engineering disciplines along with in-depth understanding of how text, graphics, audio and video can be seamlessly integrated together.

Effective development of Internet applications requires us to grasp a set of powerful techniques to plan and manage the project and monitor and control changes. In addition, implementing project management techniques will allow us to perform multiple projects in an increasingly complex environment and measure our performance improvements. Internet project management is process driven and will provide the mechanism for getting the best from our people. In this section, we describe the methodology for Internet development projects in general, and web-based business application in particular.

7.3.1 Identify Your Business Objectives and Target Audience

The first step in developing Internet projects is to identify the business goals and objectives that determine the boundaries and scope of the project. For example, the objective of a Web site development project for a local Internet mobile shop (we call it "I-Mobile" project in this chapter) can be "to provide product and sales information about our shop's mobile phones on the Web". Based on this objective, a target audience can be identified. For example, the I-Mobile's target audience can be defined as "prospective customers".

7.3.2 Analyse the Requirements and Build Domain Knowledge

Given an objective and a target audience, the next step is to create a set of requirements that turn the vision into a tangible initiative that will require resources and time. For example, the requirement list for the I-Mobile project can include:

- Display product catalogue, pictures and prices
- Retrieve detailed product information for prospective customers
- Provide online ordering and e-payment services

7.3.3 Document Your Project Plan

Assuming all the above sections have been adequately addressed and reviewed, the project manager should roll out a project plan. The project manager should develop a project plan at the beginning of the project. The emphasis should be on communication, the integrating function in project management. The project plan will formalize and document the commitments that are made. The plan should include mechanism that defines the way we want to track and control changes and improve project and product quality. A typical project plan should include the following:

- *Project Estimation* — performs an honest estimation of project effort, time and cost.
- *Project Schedule* — describes how we allocate resources along the time-line by tracking the milestones and deliverables.
- *Project Scope* — defines the extent of the work that must be done.
- *Risk Management* — identifies and monitors potential risks and develop action plans to avoid or deal with them.

7.3.4 Build the Development Team

People are the key to our success. To be effective, a project team must be organized to focus not only on getting the best people, but also on getting the best from them. The players in a software development process can be categorized as follows: Senior managers, project managers, software engineers, customers and end-users. The senior managers decide on business objectives that significantly impact the project. The project managers plan and track the progress of the project and manage the team members. The software engineers deliver the technical solutions and engineer the product or application. The customers specify the requirements and the end-users interact with the software products or applications. A successful project team should be organized in a way that capitalizes on each person's skills and abilities.

7.3.5 Review Your Current Standards and Procedures

Companies should document their project management standards and procedures and keep them as hard copy manuals that reside on the bookshelf. The valuable experience on usage and continuous improvement of the standards and procedures, lessons learned from a particular task, can be passed on to another project team. It is also necessary to identify any outages or potential conflicts caused by company initiatives, for example, ISO 9000 certification, CMMI assessments, total quantity management (TQM), and so on. All issues raised must be addressed as early as possible.

7.3.6 Identify Project Sponsors and Business Partners

The management should not only focus on building the project teams, but also channel resources to identify the project sponsors and potential business partners.

7.3.7 Adopt Just-in-Time Training Approach

A project can be severely affected either because of a lack of training or excessive training overhead. By taking a "just-in-time" training approach, in-house staff training, computer-based training and videoconference training can be provided in a "just-in-time" fashion. This approach can also accelerate the start of a project and reduce the learning curve of the project team

7.3.8 Track the Progress

There should be an action list that notes critical actions and the committed date and the person(s) responsible. This list should be updated and reviewed at a weekly meeting. This should be an operational briefing. If there are any major problems, a follow-up meeting should be arranged as soon as possible, preferably by the end of the day. All team members should meet together once a week. The management should allow genuine feedback and encourage positive suggestions.

7.3.9 Sales and Marketing

Effective sales and marketing strategies and the smooth execution of such strategies are essential to the success of the project.

7.4 Design, Implementation and Testing

The initial success of the implementation depends heavily on the effectiveness of communication and the empowerment of the implementation team. Design and development of multimedia projects go hand in hand. In this section, we highlight a number of design and implementation issues.

7.4.1 Designing User Interface

It is always a good practice to visualize the user interface before starting to implement a prototype system. If you like your Website to be visited with pleasure that will hopefully improve profitability, you should have the following as the characteristics of the Website:

- *Fast loading times and up to date contents* — the Web users are just one click away from leaving your Website. A Website should be loaded as quickly as possible before the users become impatient.
- *Excellent visual design* — coupled with quickly arranged images and text. This makes visiting your Website a sensual experience and makes people inclined to read more.
- *Attractive layout and interesting structures* — make your visitors curious about your offers and awake their inclination to buy in your shop.
- *Easy and logical user guidance* — with constant technical support. Make it easy for first time users to navigate efficiently through your offers and find the suitable items.

7.4.2 Designing the Database

I-Mobile replies heavily on database technology. It performs few calculations and is mostly designed to collect, summarize and report on data that has been collected in a very structured and stable manner. Access to and protection of the data is, therefore, our primary concern. Any access to the database should be username and password protected. The end-users are given read-only direct access only to certain elements of database.

7.4.3 Getting User Feedback

Successful businesses listen to the feedback of the customer and act on it. There should be a regular program to gather and track the users' experience. Competitive studies to ask users what they think of your competitors' Web sites versus your Web site can help to pick up best ideas and designs. Some possible strategies include:

- *Online User feedback* — create an interface in your Website to invite users for feedback and send a thank you note once feedback is received. It is the cheapest way to capture the information that the users feel important to them. It is also a good way to get a quick response on whether you have a broken link on your site.
- *Formal Surveys* — Web-based survey can be purchased from survey companies for several hundred dollars. Typically, these surveys would involve interviewing a large number of users to document profile questions and

usage information. The survey company would then send you a set of graphs and charts showing the users' responses to the questions you created.

- *Online usage profiling* — there are now new software tools that can track the pages the users view, how long they are on the page, and the written feedback and comments they provide.

7.4.4 Security

Managing the security of sensitive corporate data is always a serious concern. We should look at different perspectives including access control, database architecture, data encryption, communication layer architecture, firewalls, proxy servers, and so on. For example, the I-Mobile system must deny access to unauthorized personnel to data they have not been granted access to, such as product discount information, margin information, and so on. An ordinary user is given a uniform resource locator (URL). A privileged user is given a user name and a password for restricted access. Once a user name and a password are entered and the page then connects to the middleware to determine if a login should be allowed. Once a user is logged into the system, internal security takes over and all data is encrypted. A proxy server is also used to redirect traffic through the Web server's open port to an internal machine that is not otherwise available outside the network. This configuration allows the Web server to intercept and evaluate traffic if necessary.

7.4.5 Reliability Growth Testing

In a well-defined software development process, data should be collected throughout the software life cycle. This data usually indicates the status and progress of planning, requirement, design, coding, testing and maintenance stages of a software project. Among all this data, failure data collected during the system testing and debugging phase are most critical for reliability analysis. During this period of time, software failures are observed and corrective actions are taken to remove the faults causing the failures. Therefore, there is a reliability growth phenomenon with the progress of testing and debugging.

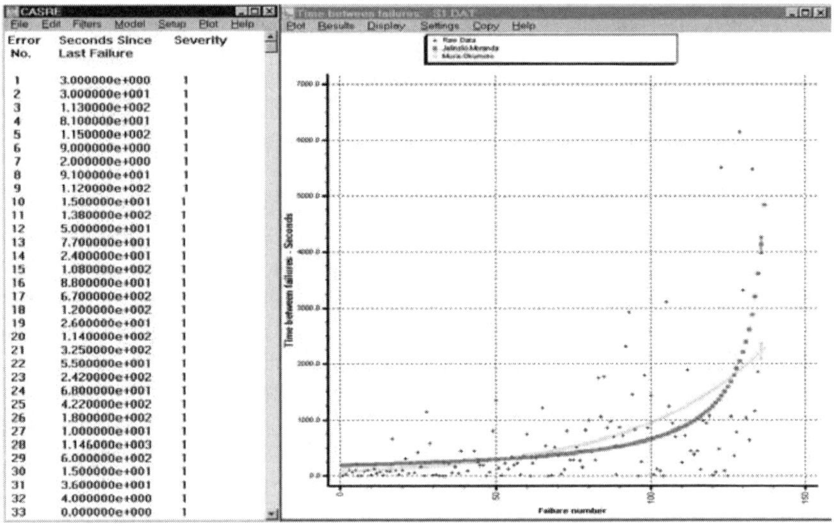

Figure 7.5 Reliability growth modelling and prediction

Appropriate reliability growth models can be used to analyse this reliability growth trend [4]. The results of this analysis are often used to decide when it is the best time to stop testing and to determine the overall reliability status of the whole system, including both hardware and software components. A good analysis will benefit a software company not only in gaining market advantages by releasing the software at the best time, but also in saving money and resources. An example of reliability growth test modelling is shown in Figure 7.5.

7.4.6 Enabling Tools and Technologies

In the past decade, there have been a number of technologies and tools developed to enable Internet and multimedia applications. For example, enabling technologies include: Java, Java Script, Active Server Pages (ASP) and Common Gateway Interface (CGI).

There are many Internet tools you may wish to acquire for your multimedia toolbox. For example, authorizing tools provide an important framework for organizing and designing your multimedia project, including graphics, sounds, animations and video clips. Web authoring tools (Dreamweaver, for example) typically rely on file transfer protocol (FTP) to update Web site files. Video editing software (VideoStudio, for example) performs functions from capturing to editing to making movies for videotape, CD or DVD or the Web. The enabling tools are categorized as the following [5]:

- Web authoring tools
- Database tools
- E-commerce tools
- Application Server tools
- Video Editing tools
- DVD Authoring tools
- Digital Imaging tools
- 3D and Animation and Web design
- Management tools

7.5. MEASUREMENTS

Why should we make measurements during a software development project? You cannot control what you cannot measure. A software metric is a unit that quantitatively determines the extent to which a software process, product or project possesses a certain attribute. In each structured software life cycle, we want to track progress, justify objectives and decisions, assess and manage risks. In this section, we first introduce a measurement methodology called the *Goal-Question-Measurement* (GQM) approach. Then we use the approach to define some measures for the Website development.

7.5.1 Identifying Metrics: Goal Question Measurement (GQM) Approach

The GQM method was first developed in the 1980s as a way to focus on the kind of data necessary to address certain perceived defects in the NASA software development process [6]. By nature, software development process is a goal driven process. The goals are predefined at the start of the project and the approaches are pre-designed towards achieving them. The GQM approach closely follows this natural thinking process. This is illustrated in Figure 7.6.

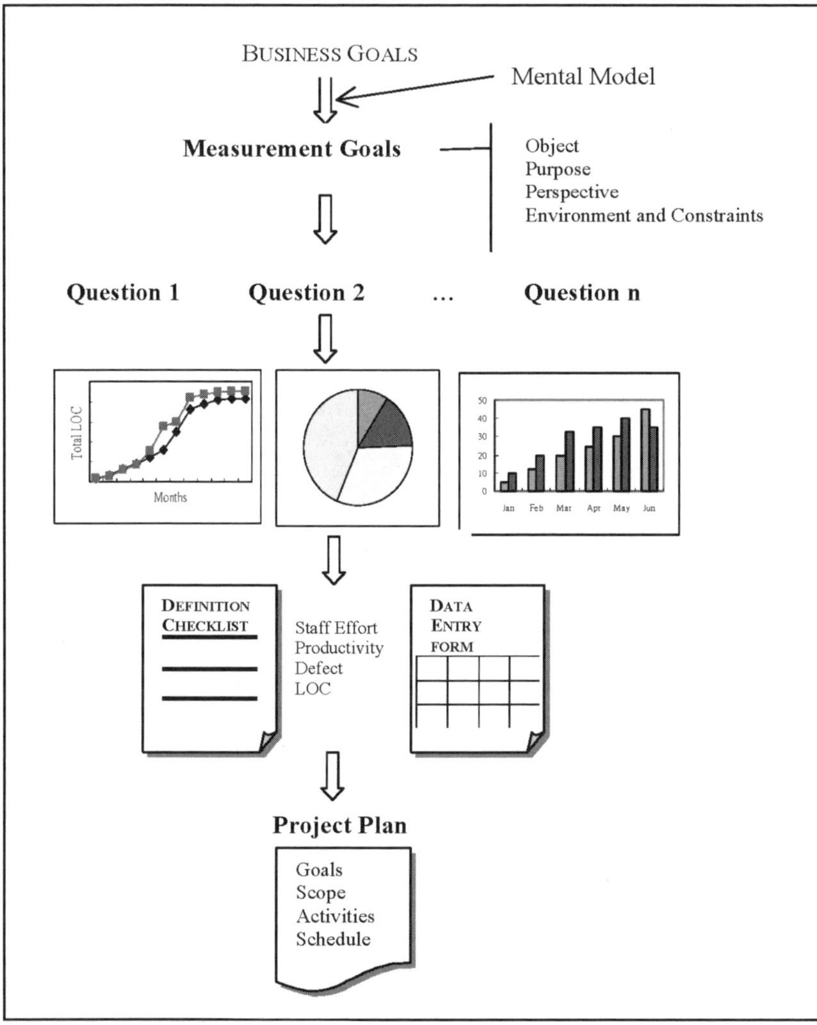

Figure 7.6 Goal-Question-Measure (GQM) approach

Since the 1980s, GQM has been widely applied to the software development process. In 1996, the Software Engineering Institute (SEI) published a well-structured guidebook on GQM [7]. The guidebook provides excellent guidance on how to lead software organizations towards achieving their business goals by determining the data needed and how to collect and measure them. The GQM measurements hierarchy contains three levels: top-level goals, questions corresponding to the goals and measurements to answer the questions. The three fundamental principles of GQM are:

- Measurement goals are derived from business goals
- The primary mechanisms for translating goals into issues, questions and measures are evolving mental models that provide context and focus
- GQM translates informal goals into executable measurement structures

7.5.2 Software Metrics

This section highlights some of the commonly used software metrics.

7.5.2.1 Schedule

A project schedule describes the software development progress by tracking its milestones completed. A project lifecycle is defined as the duration from a project start date to the formal release date of the project. The most important question asked by a project manager is whether or not the project is on time. A bulls eye chart can be used to monitor the deviations of current progress status from the plan and visualize the progress towards the target: formal project release date. An example of the bulls eye chart is shown in Figure 7.7.

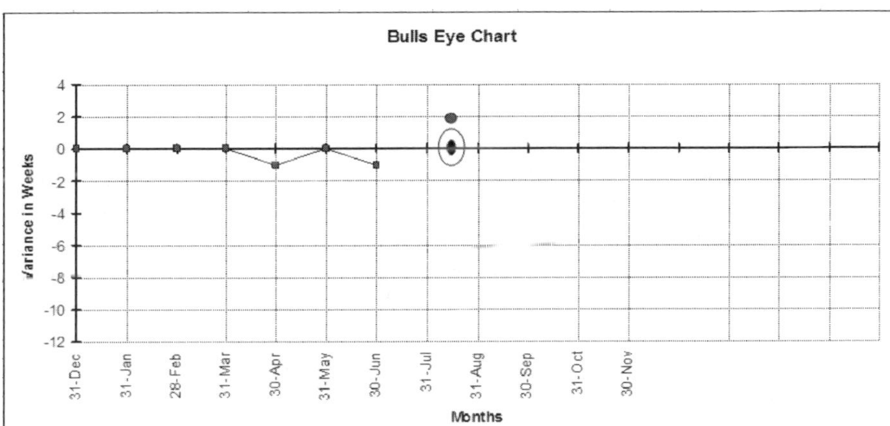

Figure 7.7 A bulls eye chart

7.5.2.2 Effort and Cost

The cost of development directly impacts the success of a project. One crucial aspect of project management is to track and monitor the cost and effort. Cost underestimation will reduce the profit margin or can even turn a profit into a loss, and cost overestimation can lead to the loss of a potential customer. A typical effort distribution measures the percentage of effort spent on a project in each development stage. An example of this is illustrated in Figure 7.8.

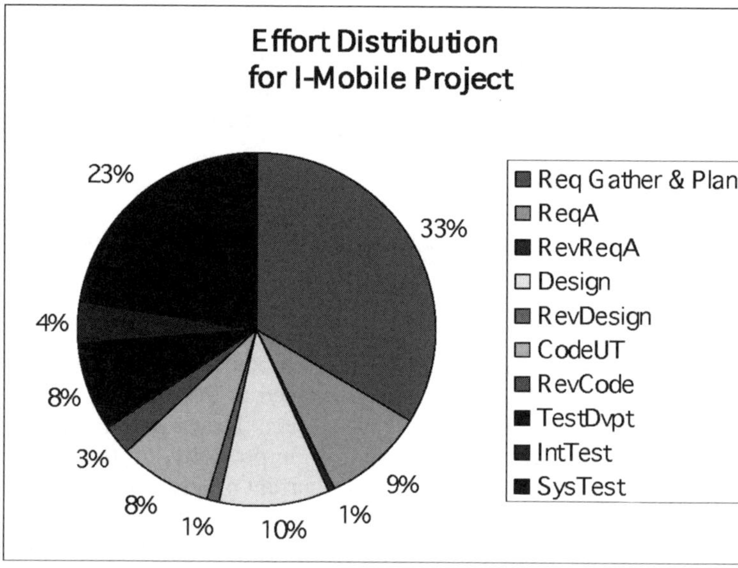

Figure 7.8 Effort distribution

7.5.2.3 Measuring Process: Trend Analysis

To effectively monitor the progress of a development project, it is also necessary to measure the process and analyse the trend. Controlling a process means keeping a process within its performance boundaries and checking whether or not the process is behaving consistently. This involves measuring the information, detecting any trend change and taking corrective actions if necessary. A typical example of a run chart for trend analysis is shown in Figure 7.9. For example, if the error density suddenly becomes increasingly high, the management investigate and determine the cause so that corrective intervening action can be taken to ensure that this will not become a long-term trend.

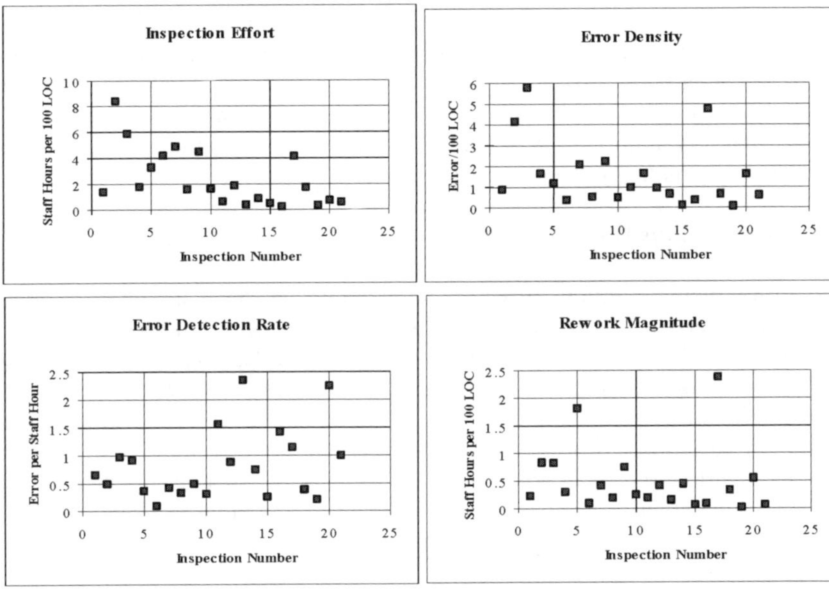

Figure 7.9 Example of using Run charts to monitor software code inspection process

7.5.2.4 Organization Level Measurement: Capability Maturity Model

In the past decade, there has been significant emphasis on improving organizational level maturity. One of the most famous is the Capability Maturity Model (CMM) proposed by the SEI [8].

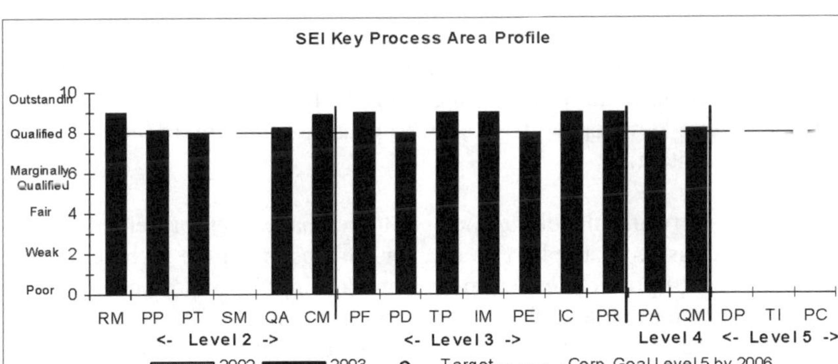

Figure 7.10 Organizational capability maturity monitor chart

The CMM serves as a guide to process improvement. It defines five maturity levels as a consequence of evaluating the key software engineering functions key Process Areas, (KPA): initial, repeatable, defined, managed and optimizing. The result can be drawn in a single chart as an overview of the organization's process capability and maturity. Figure 7.10 illustrates an example of an organizational capability maturity monitor chart.

7.5.3 Continuous Improvement

Process improvement within software organizations has been an enduring focus in both academia and industry. Many experienced managers are unsuccessful at managing software development projects because they fail to understand the cause-effect relations of the underlying software development process. Traditional project management by cost and schedule is often too late to identify problems and to take corrective actions. A program to constantly monitor and improve development process in relation to project goals is always desired by managers. Very often, continuous improvement can only be brought about with a cultural change within an organization. Every team member has a part to play, and should be entrusted to do so.

7.6 Conclusion

Advances in Internet and multimedia technologies have enabled companies and individuals to conduct business transactions online. For business or not, Web applications are primarily implemented in software. Software engineering methodologies, which have been developed and proven in the more traditional software industry, play an important part in the development of any Web-based system. These methodologies are useful not just for design and development, but also for testing and quality assurance.

This chapter adopts an all-round approach in describing the implementation of Internet and multimedia projects, including technical, managerial and business aspects of system development. It also emphasizes the importance of software process in this context.

Popular software modelling methodologies have been presented. Using the I-Mobile project as an example, this chapter has described the technical and non-technical aspects of project development, emphasizing the importance of contributions from every participant, from managers to engineers. This chapter has also presented practical guidelines for effective management and planning of projects. It has also highlighted the importance of taking meaningful measurements during the project development lifecycle. Using the data obtained, analysis can be performed to determine the status of the project: how is the project progressing? If things are not going well, what is the cause of the problem? Corrective actions can then be taken to rectify the problems in a timely manner.

These measurements allow us to perform intra-project improvements. Taking this a step further, an organization should strive to continually improve itself through inter-project improvements. The CMM has been adopted by many organizations as a vehicle for achieving improvements.

References, Links and Bibliography

[1] W.W. Royce, "Managing the development of large software systems: Concepts and techniques`", *Proc. WESCON* , Aug, 1970.
[2] B.W. Boehm, "Software Engineering Economics", Englewood Cliffs, NJ, Prentice Hall, 1988.
[3] R. Pressman, "Software Engineering: A Practitioner's Approach", 5th ed. McGraw-Hill, 2001.
[4] M.R. Lyu, "Handbook of Software Reliability Engineering", McGraw-Hill, New York, 1996.
[5] M. Dastbaz, "Designing Interactive Multimedia Systems", McGrwa-Hill, London, 2002.
[6] V.R. Basili, G. Caldiera and D. Rombach, "The Quality Question Metrics Approach", *Encyclopedia of Software Engineering,* Wiley. 1994.
[7] R.E. Park, W.B. Goether and W.A. Florac, *Goal-Driven Software Measurement – A Guidebook*, CMU/SEI-96-HB-002, Aug 1996.
[8] M. Paulk, Capability Maturity Model for Software, Software Engineering Institute, Carnegie Mellon University, Pittsburgh, PA, 1993. www.sei.cmu.edu/cmm/cmm.html

CHAPTER 8

FROM E-COMMERCE TO M-COMMERCE

8.1 ELECTRONIC COMMERCE

Electronic Commerce, or commonly referred to as e-commerce, changes the way we conduct business transactions. To facilitate e-commerce transactions handling, a common question of "what technology investments are required to succeed?" must be addressed.

Technological advancements allow more service enhancements that were previously unavailable in the traditional business model. Even before e-commerce became popular several years ago, electronic tools were widely used in dealing with various transactions such as the use of *electronic data interchange* (EDI) for sending invoices and placing orders throughout the world. Before the widespread adoption of Internet by the multitude, products and services had mostly been advertised through printed material as well as television and radio broadcast to attract telephone orders. These are examples where commerce has been supported by electronic technology for around half a century. Our goal of using technology to improve the business process has been made a reality with the Internet, and related enabling technologies such as multimedia data processing and security.

An e-commerce trading web site may be used to handle business-to-business (b2b) and/or business-to-consumer (b2c) types of e-commerce transactions. When consumers log on to your Web site and make purchases, this interface on the Internet facilitates business-to-consumer dealings. Once this sale takes place, another business-to-business event may be activated behind the scene to handle the transaction and possible dealings with another company (e.g. supplier). Another scenario is when handling online payment with a charge card, your Web site needs to interact with an external business server by various means such as Secure Socket Layer (SSL) and Secure Electronic Transaction (SET) [1]. An e-commerce Web site has to effectively deal with two parties when handling a single transaction. With the SET standards first released by Visa and MasterCard in December 1996, a secure way of making online payments for goods and services was made available to consumers worldwide [2].

Multimedia Engineering A. C. M. Fong & S. C. Hui

Truly global e-commerce is supported by additional features like multilingual trading Web sites and currency conversion tools to overcome the difficulties experienced by users based in different countries. However, one restriction with the conventional Internet e-commerce model is that it only reaches computer users who are connected online, mostly via a fixed line connection of some form. Moving from e-commerce to mobile commerce (m-commerce) allows businesses to reach different people all over the world, irrespective of their geographical location. Although many similarities exist between e-commerce and m-commerce, the latter is set to provide enhanced features with the flexible of trading anywhere irrespective of whether the user is indoors or outdoors.

8.2 Going Mobile

Global trends of multimedia service enhancements continue to fuel a strong demand for reliable and cost-effective ways of keeping people connected while on the move. The ability to conduct businesses on the move also eases the restriction on the whereabouts of customers when they trade. M-commerce is generally a description of carrying trading using a mobile device. This is to allow e-commerce made directly available to people regardless of their geographic location by using wireless communication technologies.

Enhancements in the way trading is done online makes m-commerce particularly attractive to people on the move. There are numerous examples of applications where we find m-commerce particularly appealing. M-commerce keeps you updated with the latest information that is available, regardless of where you are, as long as you are connected to the network. The most important attribute for providing m-commerce services is to maximize network availability in order to ensure that users do not miss out vital information due to service disruption. To get the most updated information, online connectivity should ideally be maintained at all times. However, this is a very challenging business because it is impossible to ensure 100% availability for wireless networks. Radio waves that are used for transferring data between the base station and end-user node propagate in a harsh environment subjected to many forms of interference and distortion. The wireless link is particularly vulnerable in an outdoor environment due to interruptions caused by uncontrollable atmospheric conditions. These factors make network availability often unpredictable. The topic of maximizing network availability in an outdoor environment is discussed later in this chapter.

Mobility finds its importance in recent years with the technological advancements of highly portable devices. Various solutions such as cellular phone systems and wireless local area networks (WLANs) provide fast and reliable wireless network access with high mobility. These systems can support a range of m-commerce functions.

Mobile shopping from virtually anywhere around the world is made possible by evolving wireless communication technologies. Trading can be completed by one of the following methods such as those provided by MasterCard International:

- *Smart Card* — the use of smart card data storage on mobile phones, personal digital assistants (PDAs) or some form of mobile computing devices.
- *Smart Card data stored in mobile phone* — similar to the above except that the smart card data is on-chip inside the mobile phone, this is particularly suitable for phones that do not have a subscriber identity module (SIM) card, for example a code division multiple access (CDMA) mobile phone.
- *Portable Wallet* — using a mobile phone to access bank accounts.
- *Mobile verification* — payment card data is verified by the mobile phone, such services can be extended to other devices such as PDAs.

While most purchases can be made online without the need for mobility, there is often up to date information that one requires to obtain while on the move. Typical examples include currency and stock exchange rates, tracking auctions, and so on. There are numerous ways of obtaining such information, such as looking over display boards in a bank. However, sourcing real-time information online, such as that illustrated in Figure 8.1, not only provides accurate information to users anywhere at any time, the user can make transactions immediately without joining a queue in the bank, or being confined to a computer with a wireline internet connection.

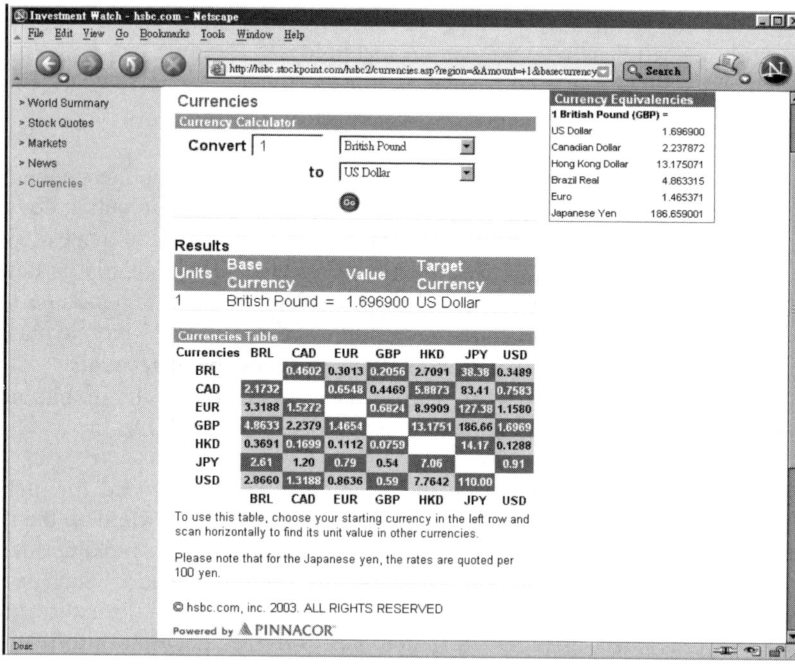

Figure 8.1 Real-time currency exchange rates conveniently displayed on the screen

As a further example to illustrate the usefulness of m-commerce, online auction has become widely accepted as a convenient way to trade for the best possible price. In order to keep track of bidding and to place a bid at anytime from anywhere, m-commerce provides a secure and reliable way to help ensure that people on the move are not missed out. Figure 8.2 shows an example of an online auction where a mobile user can get real-time outbid notice and place another bid (if desired) irrespective of one's whereabouts.

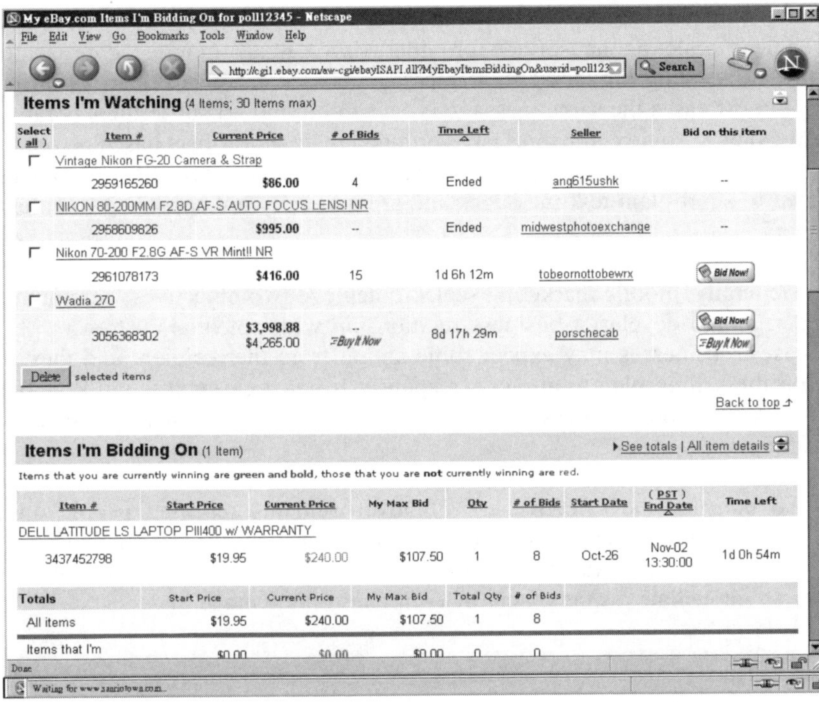

Figure 8.2 Auction over wireless access ensures bidders on the move are not missed out

In summary, m-commerce service should be provided by a wireless communication system with high availability and reliability at low costs so that anyone who has a suitable mobile device can trade while on the move. Figure 8.3 shows the block diagram of a basic system that can provide m-commerce services.

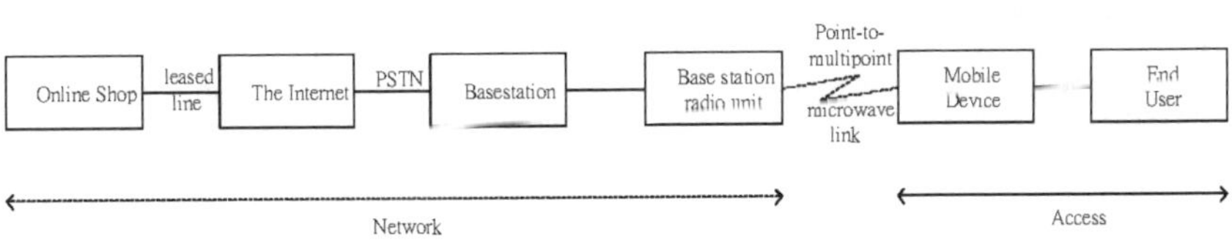

Figure 8.3 A typical m-commerce system

8.3 MARKETING AND MOBILITY

The Internet has redefined the way business communications take place. Customized promotion is made very easy by marketing through the web. Businesses can focus on marketing at an individual customer level by the delivery of classified promotional materials based on one's interests and demographic factors. Traditionally, some mobile phone service providers have made mobile advertisements available by sending broadcast short message service (SMS) to their subscribers. While in most cases the same message is sent in bulk to all subscribers regardless of the recipients, some companies have grouped their subscribers based on parameters such as gender, age group and interests. Most advertisements are generally restricted to short plain text messages and are generally not appealing to potential customers.

Generally, mobile marketing seeks to achieve two objectives: communicate the message and develop a business relationship with potential customers [3]. In some cases, businesses may expect some action from the recipient and they may focus on the selling plan to attract a response. In most cases, it is unlikely that a customer will respond unless there are some incentives and a response can be made easily.

To enhance the effectiveness of advertisements reaching people on the move colour graphics makes its way to an exciting new way of marketing with short video clips. The example shown in Figure 8.4 even provides a convenient way for an immediate response sent by recipients on the move.

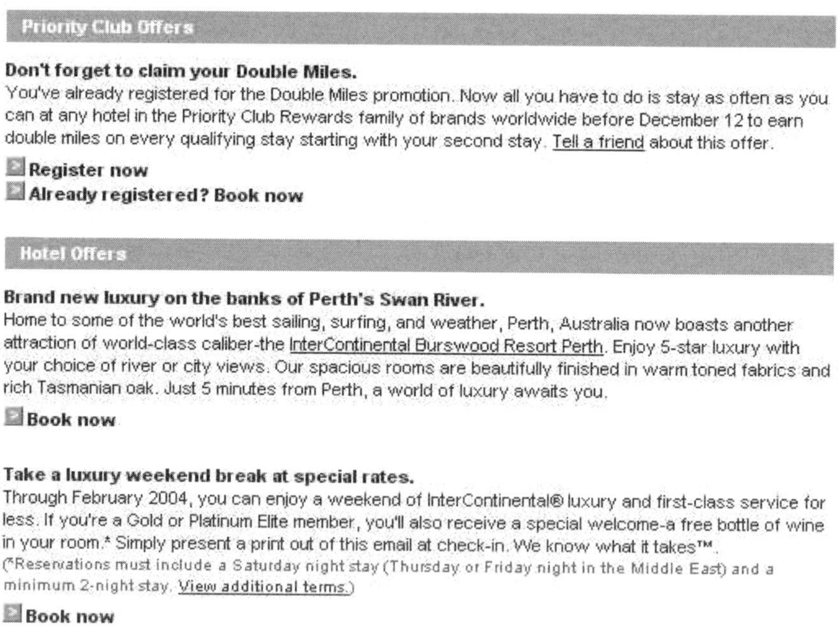

Figure 8.4 Advertisement allows a mobile user to respond

Many businesses contact their customers directly by sending email. One advantage is to let recipients simply click on a hyperlink to enter the company's web site as long as the user is connected. M-commerce makes it possible for a user to act upon receiving a message. Permission marketing, a direct marketing strategy where a customer subscribes to or agrees to receive a certain type of marketing materials, can help to maximize the chance of getting the message through to interested customers by minimizing spamming and being filtered as junk mail. At the same time, personalization can be done to match customers' interests. One good example of such is set by amazon.com where attributes related to interests such as recently looked at products are collected to suggest items of possible interest to a customer. M-commerce provides additional leverage for businesses to build individual customer's profile for effective marketing.

8.4 PROVIDING RELIABLE M-COMMERCE SERVICE IS CHALLENGING

There are many factors to consider in order to ensure that people can trade securely over wireless systems. There are many factors that determine the successful deployment of m-commerce service. In this chapter, we shall only concentrate on the three major issues that greatly affect service quality, namely, security, service availability and supported data rate.

8.4.1 Security

Mobile commerce has seen widespread adoption in the way people buy and sell, there have been numerous cases of easily exploited holes or security breaches ranging from unauthorized access to theft. Many people may be unsure about where the problems lie, and so, many potential users are deterred from trading via wireless systems due to security concerns. In this section, we look at how services can be provided with adequate security in wireless systems and how security holes can be filled.

First, let us look at networks in terms of *ad hoc* and *infrastructure* modes. *Ad hoc* networks have multiple wireless clients communicating as peers sharing data with each other with no centralized *access point*. This is the case when two persons are trading directly without dealing with an agent. On the other hand, an infrastructure network consists of several clients communicating through a central device known as *an access point*. The access point is typically a server operated by a service provider within its premises, and is usually connected to the Internet via a wired backbone.

Most m-commerce transactions take place using the infrastructure mode where a central server effectively serves as an agent. Most security measures are therefore associated with the infrastructure. For this reason, we will concentrate on security surrounding m-commerce systems in infrastructure mode. When a client is connected to the access point, its network interface card (NIC) locates the correct communication channel and establishes a session when a set of agreed protocols are established so that communication can take place. A combination of some common security features are used in m-commerce, these include service set identifier, authentication and frequency hopping.

8.4.1.1 Service Set Identifier (SSID)

Service Set Identifier (SSID) is a parameter that identifies different networks. Access points are assigned to their respective service network or service provider. The SSID is set to prevent unauthorized entry to the network. It is similar to a user password in such a way that the SSID should be made reasonably long with a combination of different character types (alphabets, numbers and symbols) so that guessing is made virtually impossible.

Each access point usually broadcasts its SSID periodically as *Beacon Frames* that let subscribers locate them to establish communication. Obviously, any person can attempt to break in with a detection tool by seeking the network name. Because of this, the network setting of SSID is only a first line of defense and its only purpose is to make it harder for intruders to find the network.

8.4.1.2 Authentication

In most cases, the first thing that occurs before a client can access the access point is to go through an *associating* process that establishes how communications take place. As soon as association is completed, authentication takes places to establish the client's identity. Authentication usually takes one of two forms, either by open authentication or key authentication.

Most systems use open authentication because it is easier to implement. In an open authentication process, any person can connect directly to the access point. This simply implies that there is no restriction on who can communicate to the access point. The only security feature here is that the access point will issue a challenge message that the client will receive and encrypt with a *Wired Equivalent Privacy* (WEP) key. The client will then send this encrypted message back to the access point for verification by determining whether there is a match.

8.4.1.3 Frequency Hopping

This is a technique that essentially switches to different channels to avoid interception. It works in such a way that the data is modulated with a carrier that switches from one frequency to another over a wide band of frequencies periodically. For example, in an IEEE 802.11b WLAN the carrier can hop between the frequency bands in the 2.4 to 2.483 GHz range. The carrier frequency is determined by a hopping code that defines the order of frequencies used. The receiver must, of course, switch to the correct frequency at precisely the same time for proper reception.

8.4.2 Reliability

The single most challenging issue about providing reliable m-commerce service is to ensure network availability by keeping the network outage time to an absolute minimum. There is a good chance of losing business when a network outage occurs, as this is equivalent to temporarily closing a shop. The Consultative Committee for International Radio (CCIR) has developed a set of procedures for the evaluation of the maximum geographical coverage and availability of wireless service. There are certain constraints to the maximum network availability as determined by parameters such as link outages and data errors [4]. A typical wireless digital communication system that provides m-commerce services usually consists of a fixed outdoor transceiver linked to the company's network backbone via wired connections. The system layout block diagram is shown in Figure 8.5. In addition to providing service to indoor subscribers, it can be seen from Figure 8.5 that service also extends to outdoor subscribers that may be several kilometers away from the outdoor transceiver's small dish antenna.

Wireless links operating in an outdoor environment are subjected to a number of atmospheric performance degradation factors such as absorption. This occurs when the propagating signal hits various air molecules, attenuation (weakening of signal strength during propagation) by water molecules due to rain, snow and fog, interference from both natural and man-made noise [5] becomes inevitable. Also, multipath fading occurs when different components of the transmitted signal is picked up by the receiver at different times due to different amounts of time delay incurred by signal components travelling in different paths to reach the receiver. These factors are described further in the following sub-sections.

8.4.2.1 Atmospheric Absorption

Gaseous absorption occurs when air molecules absorb signal energy. Different molecules have different effects on different carrier frequencies. For example, a local maximum exists with water molecules at around 22 GHz and maximum absorption by oxygen occurs at approximately 60 GHz. These are frequencies to avoid in order to minimize the weakening of signals.

8.4.2.2 Noise

Generally, noise is a degradation factor in all practical communication systems. Man-made noise is caused by a large variety of sources such as electric motors and machinery. Noise also exists in different forms naturally such as thermal noise caused by vibration of particles due to thermal energy. All these have a contamination effect added to the transmitted signal.

8.4.2.3 Multipath Fading

This is the collective effect of different components of a transmitted signal that reach the receiver at different times due to different time delays as a result of the following:

- *Diffraction* — propagating signal splits into secondary waves upon hitting an obstacle.
- *Reflection* — propagating wave hits a physical obstacle and bounces back towards the transmitter.
- *Scattering* — propagating wave hits an obstacle causing diffusion in different directions.

Figure 8.5 illustrates these phenomena as a propagating wave hits an obstacle. When the direct line-of-sight (LOS) signal path is unexpectedly blocked by obstacles, which makes the only passable signal component reach the receiver after reflection, *shadow fading* results with a noticeable reduction in received signal power.

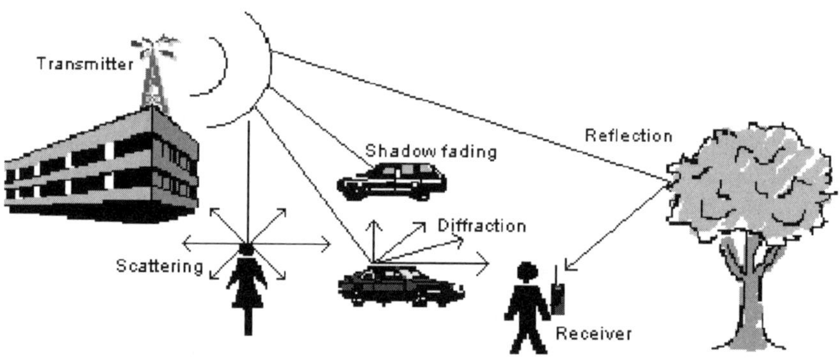

Figure 8.5 Wireless signal takes multiple paths to reach the receiver due to reflection, diffraction and scattering

Figure 8.6 In addition to multipath effects, rain causes very severe and uncontrollable disruption to wireless transmission

Microwave communication systems operating at high frequencies exhibit some undesirable propagation characteristics as free space loss increases with the square of the carrier frequency causing excessive loss. With frequencies exceeding 10 GHz, attenuation effects due to rainfall and atmospheric or gaseous absorption are large compared to other sources of signal degradation. It has been found that rain attenuation is the single most dominant source of signal degradation for wireless communication systems operating in this frequency range. Figure 8.6 illustrates the combined effects of multipath and rain on the transmitted signals. Details of the effects of rain on wireless data transmission are discussed in the next section.

8.4.3 Effects of Rain

A radio link decreases in range as rainfall becomes heavier due to more severe attenuation. Normally, a link is considered available when it can provide a bit error rate (BER) performance of at least 10^{-6}. This means that within a million bits of data transmitted across the network no more than one error bit is allowed; and a system should provide a minimum availability of 99.999%, that is, link outage allowed is just 5 minutes per year. Statistically, therefore, we assume that we only suspend trading for five minutes each year due to technical problems. Link outage caused by heavy rain is the most important factor at carrier frequencies of over several gigahertz [6]. It is therefore extremely important to ensure that the effects of rain attenuation are minimized in order to provide a sustainable network connection to provide reliable m-commerce service to users in different locations.

Rainfall has a number of undesirable effects on the transmitting signal such as attenuation, scattering and cross-polarization. Rain attenuation can cause a sharp drop in signal strength at a rate far exceeds 10 dB per km range. The *fade margin* (system operating gain) must be maintained at an adequate level to ensure that the link is still available even under persistent heavy rainfall. Generally, a higher carrier frequency and rain rate will increase attenuation significantly and any increase in data rate has very little effect on attenuation. The fade margin is usually maximized to support maximum reliability without excessive transmission power. It can be seen that the fade margin requirement increases when using more efficient modulation in terms of spectrum efficiency. The selection of modulation scheme is discussed towards the end of this chapter.

In many modern broadband wireless access (BWA) systems where frequency reuse with *orthogonal polarization* (effectively doubling channel capacity by using signals of both horizontal and vertical polarizations) is deployed to better utilize available bandwidth, cross-polarization due to rain can cause severe interference between adjacent signal paths. The effect of cross-polarization is a significant reduction in the relative phase angle between signals of vertical and horizontal polarizations as the signals propagate through rain. The extent of radio link performance degradation is measured by *cross-polarization diversity* (XPD), which is determined by the amount of coupling between signals of different polarizations. XPD typically results in a 10% reduction in coverage as a result of cell-to-cell interference.

In general, higher frequencies are affected more severely by cross-polarization and signal attenuation. According to the measurement results plotted in Figure 8.7, a difference of about 1 dB per kilometer with horizontal and vertical polarizations results when a 10 GHz carrier propagates through persistent rain of 100 mm/hr. It can be seen that horizontal polarization is much more severely affected by rain. Therefore, its relative coverage is smaller than that of vertical polarization under the same operating environments.

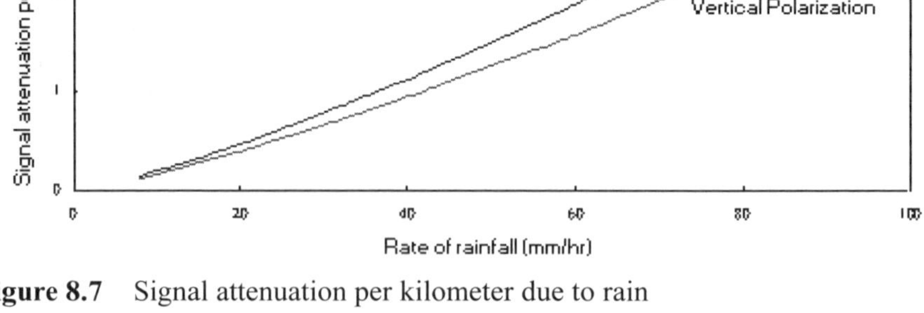

Figure 8.7 Signal attenuation per kilometer due to rain

8.4.4 Modulation Schemes

The selection of modulation schemes for m-commerce applications has identical criteria as for general mobile computing. There is always a tradeoff between spectrum utilization efficiency (SUE) versus receiver structure complexity. While multicarrier modulation such as orthogonal frequency division multiplexing (OFDM) is preferred for its relatively high degree of multipath immunity, generally low order modulation schemes such as quadrature amplitude modulation (QAM) are preferred for mobile applications mainly because of their power efficiency. Power efficiency is an important factor for mobile devices because of the need to optimize battery life and device portability.

QAM has been widely used in high-speed modems and data services for many years [7]. The maximum range, which governs the maximum distance of subscriber away from a base station, depends primarily on antenna gain and the rate of rainfall. The fade margin can be adjusted according to rainfall statistics for higher link availability to ensure that a user can remain connected almost all the time. It has been shown that systems using 16-QAM offers the best compromise among portability, power efficiency and spectral efficiency [8].

8.5 CHAPTER SUMMARY

M-commerce is becoming increasingly more popular due to its flexibility and availability. We have seen a fast growth rate in e-commerce and it is widely expected that m-commerce will take off with the current rate of technological advancements, and become an essential part of online trading. Obviously, m-commerce has numerous advantages over traditional ways of trading. However, providing secure and reliable m-commerce service is a very challenging business as discussed.

People use a variety of mobile devices for a number of transaction types. Such a diverse range of business activities is supported by a single multimedia service system. An m-commerce system must be able to provide adequate quality of service for trading anytime, anywhere. Such systems are subjected to many uncontrollable factors, both man-made and natural. Wireless communication systems operate in very harsh outdoor environment with severe impacts caused by atmospheric conditions and people movement. We tend to adjust various system parameters such as fade margin and transmission methods to maximize quality of service.

References, Links and Bibliography

[1] http://www.setco.org/set.html
[2] http://www.mastercardintl.com/newtechnology/mcommerce/

[3] W. Andrews, "E-mail marketing is stumbling forward", Internet World, pp. 13–14, March 8, 1999
[4] "Propagation in non-ionized media", Reports of the CCIR (Consultative Committee for International Radio) 718–3 1990
[5] S. Shibuya, A basic atlas of radio wave propagation, Wiley, 1987
[6] B. Fong, P.B. Rapajic, A.C.M. Fong, and G.Y. Hong, 'Polarization of received signals in a heavy rainfall region', *IEEE Commun. Lett.*, Vol. 7, No. 1, pp. 13– 14, Jan 2003
[7] G.H. Lim, D.B. Harman, G. Huang and A.V. Nguyen, "51.84 Mb/s 26-CAP ATM LAN standard", *IEEE J Sel. Areas Com.*, Vol. 13, No. 4, pp. 620–632, May 1995
[8] B. Fong, G.Y. Hong, and A.C.M. Fong, "A modulation scheme for broadband wireless access in high capacity networks", *IEEE Trans. Consumer Electr*, Vol. 48, No. 3, pp. 457–462, August 2002

APPENDIX A

POPULAR COLOUR MODELS

A.1 THE CIE CHART, RGB CUBE AND HSV SPACE

The Commission Internationale de l'Eclairage (CIE) chart defines a two-dimensional space in terms of chromaticity and luminance and is used in the Comité consultatif international pour la radio (CCIR-601-1) Recommendation Colour schemes that are based on the CIE chart, such as the Y component represents the luminance information, and is the only component used by black-and-white television receivers. I and Q represent the chrominance information. I stands for in-phase, while Q stands for quadrature (YIQ) (luma + two colour components) are mainly for colour signal transmission used in the television broadcast industry.

The Red, Green, Blue (RGB) cube, which defines a three-dimensional colour space with the three primary colours as axes, is perhaps the most widely used colour scheme for display, for example, on a computer screen. In an n-bit RGB colour scheme, a pixel may be represented by three components (Red = x, Green = y, Blue = z), where x, y and z are integers in the range $0..2^n-1$. For example, a purely red pixel is represented by the values (255,0,0) in an 8-bit RGB scheme.

The Hue, Saturation, Value (HSV) colour space represents colour in a three-dimensional space composed of two identical triangular-based pyramids (tetrahedrons) with their bases completely fused together so that one is the inverted version of the other. The HSV space is gradually gaining acceptance because it models human vision among these colour models most accurately. Figure A.1 illustrates these colour models.

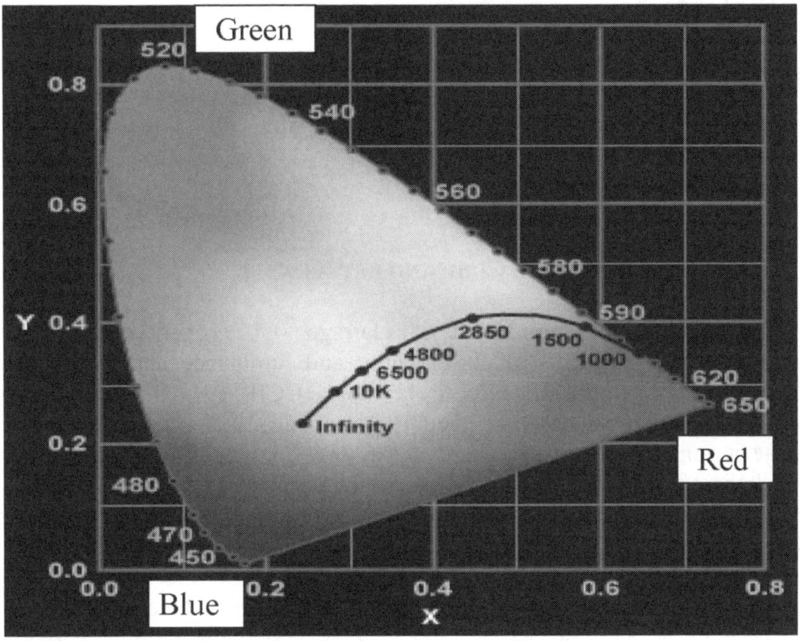
Green
520
540
560
580
590
620
650
2850
1500
1000
4800
6500
10K
Infinity
Red
480
470
450
Blue
0.8
0.6
Y 0.4
0.2
0.0
0.0
0.2
0.4
0.6
0.8
X

(a)

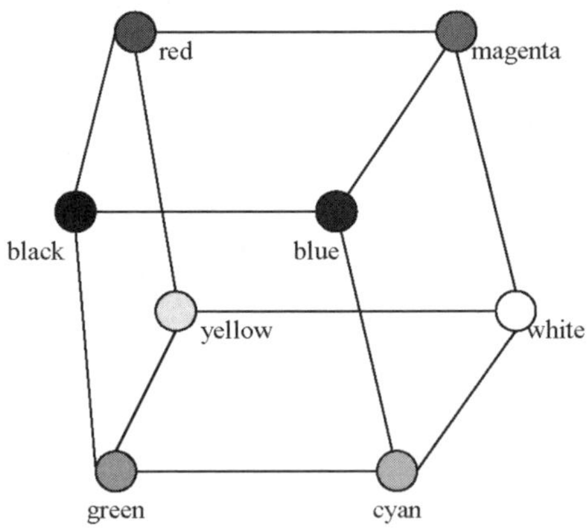
red
magenta
black
blue
yellow
white
green
cyan

(b)

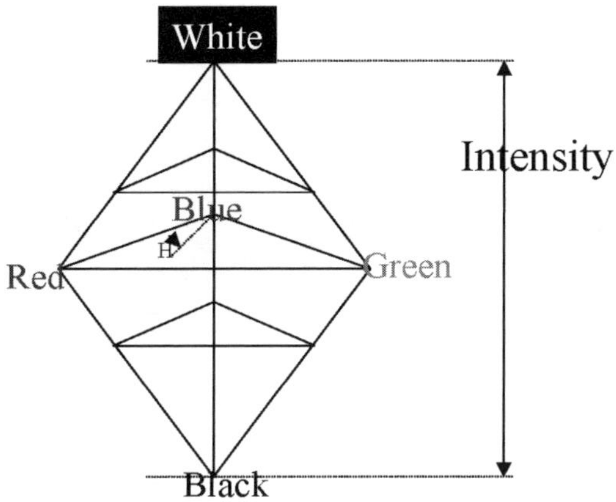

(c)

Figure A.1 Three popular colour models: (a) CIE chart, (b) RGB cube and (c) HSV space.

APPENDIX B

GLOSSARY

ADSL	Asymmetric Digital Subscriber Line
AI	Artificial intelligence
ART	Adaptive Resonance Theory, type of NN
ATM	Asynchronous transfer mode (cell-switched)
B2B	Business-to-business (trading)
BER	Bit error rate
BWA	Broadband wireless access
CBR	Case-based reasoning
CCITT	Comite Consultatif Internationale de Telegraphie et Telephonie (now ITU-T)
CMM	Capability Maturity Model
Codec	Coder-decoder
DBNN	Decision-based neural network
DCT	Discrete cosine transform
DiffServ	Differentiated services
DSL	Digital subscriber line
DWT	Discrete wavelet transform
EDI	Electronic data interchange
FTP	File transfer protocol
Fuzzy ART	Fuzzy version of Adaptive Resonance Theory NN
G.7xx	ITU audio codecs for H.32x videoconferencing
GSTN	General switched telephone network
H.26x	ITU video codecs for H.32x videoconferencing; bit rate can be low enough for V.34 modem using 20 kbps for video and 6.5kbps for audio. H.236 is an extended version of H.261 with more format supported and motion prediction.
H.32x	H.320 ITU standard for videoconferencing over ISDN H.323 ITU standard for videoconferencing over packet network (LAN) H.324 similar to H.323, but optimized for low bit rate transmissions over PSTN
HITS	Hyperlink-induced topic search

HTTP	Hypertext transfer protocol
IETF	Internet engineering taskforce
IntServ	Integrated services (Internet)
IP	Internet protocol
ISDN	Integrated services digital network
ISM	Intelligent Scene Monitoring
ISO	International Standards Organization
ITU	International Telecommunications Union
ITU-T	ITU- Telecommunications Standardization Sector
JPEG	Joint photographic experts group
LAN	Local area network
LOS	Line of sight
LVQ	Learning vector quantization, type of NN
MCU	Multipoint control unit
MPEG	Moving picture experts group
NN	Neural network
OFDM	Orthogonal frequency division multiplexing
OLAP	Online analytical processing
P2P	Peer-to-peer
PC	Personal computer
PICS	Platform for Internet Content Selection
POTS	Plain old telephone service
PSTN	Public switched telephone network
QAM	Quadrature amplitude modulation
QoS	Quality of service
RFC	Request for comment (an approved IETF document)
RSA	Ron Rivest, Adi Shamir and Leonard Adleman, Inventors of a popular public key cryptography scheme
RSVP	ReSerVation set-up protocol
RTCP	Real-time control protocol
RTP	Real-time transport protocol
RTSP	Real-time streaming protocol
SEI	Software Engineering Institute
SET	Secure electronic transaction
SHTTP	Secure HTTP
SIP	Session initiation protocol
SMS	Short message service
SOM	Self-organizing map
SSID	Service set identifier
SSL	Secure sockets layer
STM	Synchronous Transfer Mode
SUE	Spectrum utilization efficiency

T.120	ITU standard that is made up of a suite of communication and application protocols that facilitate data conferencing features, such as program sharing, whiteboard conferencing and file transfer. T.120 can be combined with other ITU standards, such as H.32x for audio and video conferencing.
TCP	Transmission control protocol
TDM	Time-division multiplexing
TQM	Total quality management
URI	Uniform resource identifier
URL	Uniform resource locator
UDP	User datagram protocol
V.34	ITU standard modem serial line protocol that can support symbol rates up to 28,800 bps
VMD	Video Motion Detection
VoIP	Voice over IP
WWW	World Wide Web (often shortened to just "Web")
XPD	Cross polarization diversity

INDEX